This book series discusses the various infectious diseases affecting the livestock, principle of the disease control, and specific disease management. It discusses the existing strategies to control infectious disease includes animal management programs, vaccination, targeted antimicrobial use, and food hygiene.

Despite public health and veterinary public health improvement within the last century, animal populations remain vulnerable to health threats caused by infectious diseases.

It reviews the current understanding of the zoonotic, emerging, and transboundary animal infections in relation to their transmission, epidemiology, clinical and pathological effects, diagnosis and treatment.

It also examines the recent advancements in the veterinary diagnostics including the existing capabilities, constraints, opportunities, and future potentials. In addition, it elaborates on the conventional and recombinant vaccines that are used in the veterinary medicines and the molecular approaches that have led to the development of new vaccines in recent years. A volume focusing on the various water- and food-borne diseases and its impact on the domestic animals is also a part of this series.

The book series examines the emergence of antimicrobial resistance in livestock, ongoing global surveillance, and monitoring program, its impact on the animal-human interface and strategies for combating resistance.

Livestock Diseases and Management

Series Editor
Yashpal Singh Malik, Guru Angad Dev Veterinary and Animal Sciences University College of Animal Biotechnology, Ludhiana, Punjab, India

Editorial Board
Rameshwar Singh, Bihar Animal Sciences University, Patna, Bihar, India
A. K. Gehlot, Rajasthan University of Veterinary & Animal Sciences, Bikaner, Rajasthan, India
G. Dhinakar Raj, Centre for Animal Health Studies, Tamil Nadu Veterinary and Animal Sciences, Chennai, Tamil Nadu, India
K. M. Bujarbaruah, Assam Agricultural University, Jorhat, Assam, India
Sagar M. Goyal, Institute of Molecular Virology, University of St. Thomas - Minnesota, Saint Paul, MN, MN, USA
Suresh K. Tikoo, School of Public Health, University of Saskatchewan, Saskatoon, SK, Canada

Tanmoy Rana • Benito Soto-Blanco
Editors

Good Practices and Principles in Pig Farming

Editors
Tanmoy Rana
Department of Veterinary Clinical Complex
West Bengal University of Animal and
Fishery Sciences
Kolkata, West Bengal, India

Benito Soto-Blanco
Veterinary School
Universidade Federal de Minas Gerais
Belo Horizonte, Minas Gerais, Brazil

ISSN 2662-4346　　　　　　　　ISSN 2662-4354　(electronic)
Livestock Diseases and Management
ISBN 978-981-97-4664-4　　　　ISBN 978-981-97-4665-1　(eBook)
https://doi.org/10.1007/978-981-97-4665-1

© The Editor(s) (if applicable) and The Author(s), under exclusive license to Springer Nature Singapore Pte Ltd. 2024
This work is subject to copyright. All rights are solely and exclusively licensed by the Publisher, whether the whole or part of the material is concerned, specifically the rights of translation, reprinting, reuse of illustrations, recitation, broadcasting, reproduction on microfilms or in any other physical way, and transmission or information storage and retrieval, electronic adaptation, computer software, or by similar or dissimilar methodology now known or hereafter developed.
The use of general descriptive names, registered names, trademarks, service marks, etc. in this publication does not imply, even in the absence of a specific statement, that such names are exempt from the relevant protective laws and regulations and therefore free for general use.
The publisher, the authors and the editors are safe to assume that the advice and information in this book are believed to be true and accurate at the date of publication. Neither the publisher nor the authors or the editors give a warranty, expressed or implied, with respect to the material contained herein or for any errors or omissions that may have been made. The publisher remains neutral with regard to jurisdictional claims in published maps and institutional affiliations.

This Springer imprint is published by the registered company Springer Nature Singapore Pte Ltd.
The registered company address is: 152 Beach Road, #21-01/04 Gateway East, Singapore 189721, Singapore

If disposing of this product, please recycle the paper.

Preface

The good practices and principles in pig farming generally constitute the main livelihood of rural poor belonging to the lowest socioeconomic conditions, aiming to educate scientific pig farming for maintaining improved foundation stock, proper housing, feeding, management, and economic upliftment. Therefore, suitable protocols or implementation strategies should be maintained for the popularization of enhancement of scientific knowledge of pig breeding cum rearing of meat-producing animals with adequate financial provisions. The good practices are necessary to modernize pig farming for improved productivity in small-sized rural pig farms in generating an economy for the small and marginal farmers. The good practices aim to secure food and nutritional security for a growing population with a need for an integrated approach to pig farming. Pig farming is the most potential source of pork production, bristles, and manure. It also generates employment opportunities for rural farmers to improve their living standards. This book constitutes several chapters and also provides new thoughts and scientific modules for the betterment of understanding of the good practices of pig farming. This book aims to maintain empowerment and improvement of the pork production chain with a piece of proper basic knowledge or support to the daily activities of medium size pig farming management. This book helps students, agricultural academicians, advisors, animal scientists, veterinary practitioners, progressive farmers/sheep owners, and farm managers intending to provide the best available supervision for improving the productivity of pigs.

Kolkata, West Bengal, India	Tanmoy Rana
Belo Horizonte, Minas Gerais, Brazil	Benito Soto-Blanco

Acknowledgement

I would like to convey my sincere gratitude to the Dean of the Faculty of Veterinary and Animal Sciences and Hon'ble Vice Chancellor, West Bengal University of Animal & Fishery Sciences, Kolkata, India, for assisting in editing this book. I am also extremely grateful to all the contributors who contributed to this book within the stipulated time. I would like to express my warmest thanks to the departmental colleagues who provided me with extreme energy for editing this book. Finally, I would also like to express my sincere gratitude to Dr. Naren Aggarwal, Editorial Director Medicine, Biomedical and Life Sciences Books Asia; Swati Sharma and Soumya Basu, Associate Editors; Muthuneela Muthukumar, Production Editor, Springer Nature; and other members who actively or indirectly helped me to edit this book and provided an opportunity for the publication of this book. This book is dedicated especially to veterinary students, researchers, academicians, farm managers, veterinary practitioners, and progressive farmers for the enhancement of knowledge of the farm. I hope this book will be a very useful resource for the readership.

Contents

1 **Introduction** 1
   Amitava Roy, Tanmoy Rana, and Partha Sarathi Roy

2 **Breeds of Pigs of India and its Productivity** 15
   N. Rajanna, J. Saikiran, and J. Shashank

3 **Indigenous Pigs Breeds of Nepal** 31
   Subir Singh

4 **Body Condition Scoring of Pigs** 41
   Vallabhaneni Srikanth and Cherryl Dimphna Miranda

5 **Pig Production and Livelihood Security** 57
   Saroj K. Rajak, Satish Kumar, Jaya Bharati, Anil Kumar, Kumar
   Shambhu Sharnam, and Divya Rani

6 **Pig Behavior and Welfare** 77
   Subir Singh

7 **Pig Exhibition Rules, and Its Monitoring** 85
   Jessy Bagh, Annada Das, Kaushik Satyaprakash, and Tanmoy Rana

8 **Impact of Mycotoxins on Pig Production** 105
   Amitava Roy and Tanmoy Rana

9 **Effect of Environment on Pig's Health** 123
   Amitava Roy and Tanmoy Rana

10 **Production of Biofuel from Pork Fat** 141
    Felix Uchenna Samuel and Jacob Oluwoye

11 **Pig Farming and Business Opportunities for Financial Benefit** 171
    Saroj K. Rajak, Jaya Bharati, Satish Kumar, Rakhi Bharti, and
    Pinky Preety

12 **Entrepreneur Development Through Pig Farming** 189
    S. Swetha Kanthi and Srikanth Vallabaneni

Editors and Contributors

About the Editors

Tanmoy Rana is currently working as an Assistant Professor of Veterinary Clinical Complex at West Bengal University of Animal & Fishery Sciences, Kolkata, India. He has a bachelor's degree in veterinary sciences and animal husbandry, a master's degree in veterinary medicine and ethics and jurisprudence from West Bengal University of Animal & Fishery Sciences, Kolkata, India, and a Ph.D. in veterinary science from the University of Calcutta, Kolkata, India. His research interests involve arsenic toxicity, molecular diagnosis, molecular toxicology and medicine, oxidative stress, immunopathology, nanoparticles, echinococcosis, and host–microbe interactions. He is actively engaged in teaching and clinical practices in veterinary medicine and research related to animal health, production, and disease monitoring regimes. He has published several research articles in reputed international and national journals along with review articles in international journals. He is an editorial board member of several journals and has published various national and international books.

Benito Soto-Blanco is currently a Professor of the Department of Veterinary Clinic and Surgery at the Veterinary School, Federal University of Minas Gerais, Brazil. He is a highly accomplished veterinarian with a bachelor's degree in veterinary medicine from Universidade Paulista, Brazil and a master's and Ph.D. in experimental and comparative pathology from Universidade de São Paulo, Brazil, with a period at ARS/USDA Poisonous Plants Research Laboratory, Logan, UT, USA. He has focused his research on veterinary toxicology, pathology, and clinical pathology. He has published more than 200 research articles in peer-reviewed international journals and authored or coauthored two books and more than 20 book chapters. He is Editor-in-Chief of the Brazilian Journal of Veterinary Pathology.

Contributors

Jessy Bagh Department of Livestock Production and Management, College of Veterinary Sciences and Animal Husbandry, OUAT, Bhubaneswar, Odisha, India

Jaya Bharati ICAR—National Research Centre on Pig, Guwahati, Assam, India

Rakhi Bharti Department of Veterinary and A.H. Extension Education, CoVAS, Kishanganj, Bihar, India

Annada Das Department of Livestock Products technology, Faculty of Veterinary and Animal Sciences, WBUAFS, Kolkata, West Bengal, India

S. Swetha Kanthi Department of Veterinary and A.H Extension Education, Sri Venkateswara Veterinary University, College of Veterinary Science, Proddatur, Andhra Pradesh, India

Anil Kumar Department of Veterinary Clinical Complex, Bihar Veterinary College, Patna, Bihar, India

Satish Kumar ICAR—National Research Centre on Pig, Guwahati, Assam, India

Cherryl Dimphna Miranda Department of Livestock Production Management, College of Veterinary Science, Tirupati, Sri Venkateswara Veterinary University, Tirupati, Andhra Pradesh, India

Jacob Oluwoye Departments of Urban and Regional Planning, Alabama A&M University, Huntsville, AL, USA

Pinky Preety Department of Veterinary and A.H. Extension Education, GADVASU, Ludhiana, Punjab, India

Saroj K. Rajak Department of Veterinary and A.H. Extension Education, Bihar Veterinary College, Patna, Bihar, India

N. Rajanna ICAR-Krishi Vigyan Kendra, Mamnoor, PVNRTVU, Warangal, Telangana, India

Tanmoy Rana Department of Veterinary Clinical Complex, West Bengal University of Animal and Fishery Sciences, Kolkata, West Bengal, India

Divya Rani Department of Veterinary and A.H. Extension Education, Bihar Veterinary College, Patna, Bihar, India

Amitava Roy Department of Livestock Farm Complex, West Bengal University of Animal and Fishery Sciences, Kolkata, West Bengal, India

Partha Sarathi Roy Murshidabad Krishi Vigyan Kndra, Milebasa, Bhagwangola, West Bengal, India

J. Saikiran ICAR-Krishi Vigyan Kendra, Mamnoor, PVNRTVU, Warangal, Telangana, India

Felix Uchenna Samuel Animal Science Program, Alabama Cooperative Extension, Alabama A&M University, Huntsville, AL, USA

Kaushik Satyaprakash Department of Veterinary Public Health and Epidemiology, Faculty of Veterinary and Animal Sciences, RGSC, BHU, Mirzapur, Uttar Pradesh, India

Kumar Shambhu Sharnam Animal and Fisheries Resources Department, Govt. of Bihar, Patna, Bihar, India

J. Shashank ICAR-Krishi Vigyan Kendra, Mamnoor, PVNRTVU, Warangal, Telangana, India

Subir Singh Department of Veterinary Medicine and Public Health, Faculty of Animal Science, Veterinary Science and Fisheries, Agriculture and Forestry University, Rampur, Chitwan, Nepal

Vallabhaneni Srikanth Livestock Production and Management Section, ICAR-Indian Veterinary Research Institute (ICAR-IVRI), Izatnagar, Bareilly, Uttar Pradesh, India

Srikanth Vallabaneni Division of Livestock Production and Management, Indian Veterinary Research Institute, Izatnagar, Bareilly, Uttar Pradesh, India

Introduction

Amitava Roy, Tanmoy Rana, and Partha Sarathi Roy

Abstract

The accumulation of androstenone, skatole, and other indoles in fat is what causes boar taint; this is controlled by the ratio of these compounds' synthesis to degradation and can be influenced by a variety of factors such as environment and management techniques, nutrition, genetics, and sexual maturity. Immunocastration is one method of controlling boar taint, however it is not widely used in all nations. By using genome editing or selective breeding, genetics provides a long-term solution to the boar taint issue. There are several short-term methods that have been suggested to manage boar taint, but their effectiveness varies, and there is too much variation throughout breeds and individuals to apply a general remedy. As a result, we suggest using precision livestock management to create taint control strategies. This entails identifying the genetic variants and differences in metabolic processes that result in boar taint in particular groups of pigs, and using this knowledge to develop specialised treatments based on the underlying cause of boar taint. For certain animal populations, boar taint treatments can subsequently be identified and put into practice using genetic, proteomic, or metabolomic screening. Globally, the technology utilised in the breeding and production of pigs has changed significantly in the recent past. Pig production must be raised in order to reduce employment, and

A. Roy
Department of Livestock Farm Complex, West Bengal University of Animal and Fishery Sciences, Kolkata, West Bengal, India

T. Rana (✉)
Department of Veterinary Clinical Complex, West Bengal University of Animal and Fishery Sciences, Kolkata, West Bengal, India

P. S. Roy
Murshidabad Krishi Vigyan Kndra, Milebasa, Bhagwangola, West Bengal, India

© The Author(s), under exclusive license to Springer Nature Singapore Pte Ltd. 2024
T. Rana, B. Soto-Blanco (eds.), *Good Practices and Principles in Pig Farming*, Livestock Diseases and Management,
https://doi.org/10.1007/978-981-97-4665-1_1

this can be achieved through improved herd management. Raising pigs yields significant profits since they have the best feed conversion ratio of any animal used to produce meat, with the exception of broilers. They give birth to ten piglets on average every farrowing and have shorter gestation periods—114 days on average. Even though it only takes 7 months to reach marketable weight, pig farming makes quick money. Pork, sausages, ham, bacon, and other products made from pigs are in great demand. Pigs are raised in large, integrated pig operations subject to strict biosecurity rules, as well as in backyard systems and on waste belts to family-run farms. Ensuring the welfare and health of animals is the primary measure towards accomplishing this objective. The economics of the farm's produce depend on the health of the pigs. Reducing long-term drug usage will contribute to keeping pig herds healthy and slowing the spread of numerous diseases. When dietary lysine and energy are supplied close to requirements, milk protein synthesis is enhanced and muscle protein mobilisation is reduced. Nursing sows' needs for amino acids vary because of the dynamic mobilisation of body tissues during lactation; however, lysine (Lys) is always the first-limiting amino acid. Using an early warning system, a veterinarian or herd management may determine a distinct diagnosis based on continuing observation of behavioural factors and a specific metric (such body temperature). Ensuring the welfare and health of animals is a crucial step towards accomplishing this objective. An important factor in production economics is also the health of the pigs on the farm. A widely accepted high potential herd management strategy calls for maintaining a good health status of pig herds by decreasing the incidence of various disease units and lowering the use for antimicrobial drugs.

Keywords

Performance · Herd management · Technology · Pig · Welfare

1.1 Introduction

Pork is regarded as a major commercial product worldwide, and pig rearing is an important part of the global livestock industry. The Food and Agriculture Organisation (FAO) estimates that 97 million metric tonnes of pork would be produced worldwide in 2020. China was the world's largest producer of pork, making up over half of the total. The European Union (EU), the United States, Brazil, Russia, Vietnam, and Canada are among other important pig-farming nations. Asia produces around half of the world's pigs, with the USSR and Europe producing the remaining 30%. Pork is exported by nations having over 10 years of production to satisfy demand in other areas. These exporting nations use intensive, industrial-scale pig farming practices in an effort to supply the increasing demand for pork while maximising efficiency. China, Japan, Mexico, and South Korea are among the nations that import pork goods, while the EU, the USA, and Brazil are among the top exporters of pork products. In underdeveloped countries, the most common way of raising pigs is the severe indoor confinement system. The intense indoor production technique has been criticised from the start for possible environmental harm that could endanger sustainability in Asian and European countries (Dawkins 2017). Outdoor production techniques, on the other hand, can promote animal welfare, offer up new commercial opportunities for small, resource-constrained farms, and solve both environmental and food safety issues. Due to its comparatively low investment cost and potential for additional value, outdoor production is a great alternative to small farmers' indoor confinement systems (Miao et al. 2004). Modern sows normally wean 33–35 piglets per sow per year, but the most productive herds currently wean over 40 piglets per sow annually (Krogh et al. 2016). Pigs born from contemporary sows are less vulnerable and have lower development potential as a result. The increased ability for reproduction puts the sow under increased metabolic stress throughout breastfeeding and pregnancy. Compared to their forebears, females with the current genotype grow faster and have less fat tissue. According to Thomas et al. (2018), sows in commercial production at the age of 2 and older often have fat depths between 12 and 16 mm at farrowing. At farrowing, the average girth tenth rib fat depth is 16 mm. These modifications in body composition and reproductive efficiency affect the nutritional needs for lactation and pregnancy. Pig breeds raised for growth and finishing are genetically designed to yield lean meat and high feed efficiency; however, this also affects the physiology, productivity, and feed efficiency of the reproductive females. Feeding the reproductive females is a discipline that is very different from feeding growing pigs because in female reproductive pigs, traits like adequate body fat and colostrum production, as well as optimal farrowing and lactation performance, are far more important than the traditional traits in growing pigs, where the focus is on maximising gain and feed efficiency (Kim et al. 2015). An overview of the physiological characteristics of growing gilts and reproductive sows during the phases of gestation, transition, and lactation is provided in this work, along with an update on the current knowledge of energy and lysine requirements.

1.2 Record Systems

Adoption of advanced record-keeping programmes and benchmarking has been one of the main forces behind increasing efficiency and productivity of hog production. These record-keeping programmes not only established output standards for individual farms, but they also standardised production parameters, enabling comparisons between different farms. Large datasets may therefore be created for benchmarking and to assist producers in finding areas where their herds could be improved (Ketchem and Rix 2013). Programmes for keeping records and standards for pigs reared for finishing and nursery soon followed. Key profitability factors started to be discovered thanks to the abundance of production data that was available, and the adoption of new technology could be tracked and assessed (Baas 1996). Production data is now combined with profit forecasts and financial analysis to assist reduce expenses, manage risk, and boost income. Since it is hard to improve what cannot be measured, a large portion of the increase in swine productivity most likely would not have been achievable without record keeping and benchmarking.

1.3 Selection and Breeding Management

Because of their higher meat quality, disease resistance, and quicker growth rate, a number of cross-bred pigs have also been produced and are deemed to be more suited. Along with a number of other considerations, consumer preferences and growth rate play a major role in breed selection. The next step is breeding, which is done randomly in native pigs via a scavenging strategy. This reduces their productivity and even increases their susceptibility to illnesses. Domesticated sows have a 21-day oestrus cycle and are polyestrous, meaning they are not seasonal. Successful breeding depends on oestrous identification, which calls for perceptive experts who fully comprehend the cyclic behaviour of females. Oestrous symptoms include hunger loss, restlessness, elevated vocalisation, erect ears, alert demeanour, swollen and reddish vulva, lordosis, stiffness when back pressured (standing reflex), etc. Breeding healthy sows during or after third oestrous is always recommended. Pigs have two main types of mating systems: artificial and natural. Pen and hand mating are examples of natural mating. While in pen mating, boars are let to stay with the sow herd, in hand mating, boars are transported to sow under supervision throughout each service. Artificial insemination is a different kind of mating that involves depositing semen into the female reproductive canal. In non-slippery surfaces, the optimal time for AI is 15–24 h following the commencement of oestrous.

1.4 Reproduction

The last 30 years have seen significant advancements in genetics, thanks to technological advancements in reproductive efficiency. Swine farms used pen mating in the 1980s, when boars and sows were merely placed outside on dirt lots together. It was generally advised that one boar was required for every 20 sows when it came to pen mating. Seasonal infertility greatly hindered reproductive success, and pen mating caused in wide variance in farrowing dates. Artificial insemination (AI) was the technology that revolutionised all of this, and the swine industry immediately embraced it in the 1990s (Bortolozzo et al. 2015). At first, farms would gather their own boars, but as the industry became more specialised, boar studs were developed to enable the wider adoption of high-indexing boars as well as improved methods for semen collection and extenders (Table 1.1).

Boar studs often yield 15–20 dosages of semen from a single collection these days. One boar for every 250 sows, or roughly 50,000 sows, can be fed using the semen produced by a 200-head boar stud. Because fewer boars were needed, more of the highest-indexing boars could be chosen and used, which quickly improved growth performance and reproductive efficiency. By preventing the admission of live animals—the main way that diseases are introduced to farms—bringing the

1 Introduction

Table 1.1 Reproductive quality metrics of gilt

S. no.	Parameters	Characteristic
1.	Length of heat period	2–3 days
2.	Best time to breed in heat period	Gilts—First day and sows—Second day
3.	Age to breed gilts	8–10 months
4.	Gestation period	114 days
5.	Number of services per sow	2 services at an interval of 12–14 h
6.	Weight of breed gilts	100–120 kg
7.	Onset of heat after weaning	2–10 days
8.	Period of oestrous cycle	18–24 days (average 21 days)

Fig. 1.1 Coupling of pig

male through AI significantly decreased the likelihood of disease. In the past, artificial insemination (AI) meant injecting semen into the female's cervix. However, post-cervical insemination (PCAI), in which semen is injected directly into the uterus, became accessible and started to be utilised in the industry in the 2000s (Fig. 1.1).

Less time is needed for each insemination using PCAI, but the biggest benefit is the potential for fewer sperm cells to be needed for each insemination. According to a review by Bortolozzo et al. (2015), PCAI offers excellent potential to reduce the number of sperm cells per breeding dose from 3 billion to roughly 1.5 billion, without significantly reducing the efficiency of female reproduction. This, in turn, would enable the use of higher genetically indexing boars for longer periods of time and require fewer boars to meet breeding requirements. Genetic development will soon include technologies like sexed semen, fixed-timed insemination, and the ability to transplant elite boar stem cells into sterilised recipient boars.

1.5 Housing Management

Buildings, machinery, disease prevention measures, and managerial attention are all necessary for pig farming. It is first necessary to build a modest shed that can be subsequently expanded and has a constant supply of electricity and water. There is also very little labour required. Without an assistant, the farmer and his family can manage a small farm quite successfully on their own (Fig. 1.2).

But the actual number of workers needed for a commercial farming operation would vary depending on how big the piggery is. In a large commercial farm, labourers may be required for various tasks such as feeding the pigs, washing the pigs, and cleaning and sanitising the pig housings. A farmer can successfully raise 85% or more of all live-born piglets to market weight in the lowest amount of time by adopting good housing, which facilitates management. Pigs require varying surroundings (temperatures) depending on their maturation stage. Piglets require specific protection against extremely low temperatures if they are to develop and produce to the fullest extent possible. Pigs that are growing and breeding need to be shielded from extreme heat (Fig. 1.3).

For this reason, the houses have to be constructed so that the pigs are shielded from harsh weather, including freezing winds and nonstop rain (Fig. 1.4).

Fig. 1.2 Housing with appropriate drainage

1 Introduction

Fig. 1.3 Improved pig housing

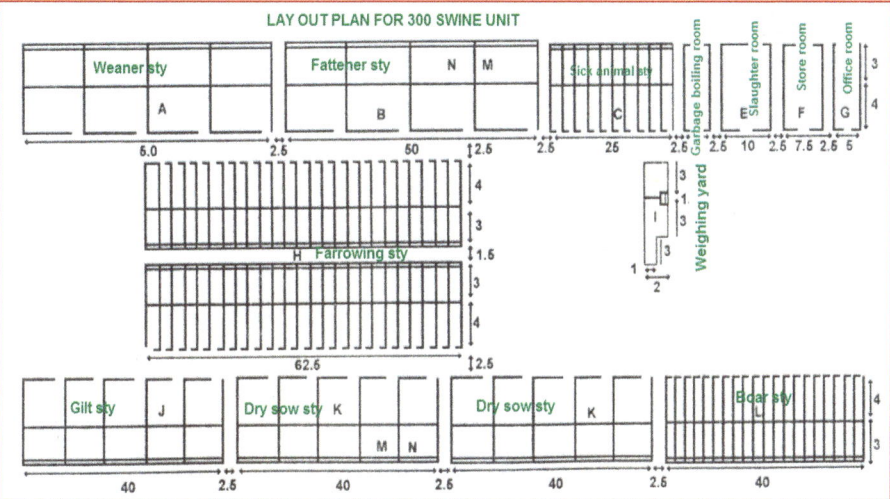

Fig. 1.4 Outline of pig sty

1.6 Care and Management of Different Categories

On farms, about half of the pigs that pass away do so before they turn 14 days old. It is consequently crucial to have excellent management in the farrowing house, where the piglets are born and raised for the first 28–35 days of their existence. As soon as possible after birth, make sure all piglets suckle a teat to receive colostrum (Fig. 1.5).

Colostrum is the first milk the sow produces right after the piglets are born. During the first few weeks of their life, the colostrum strengthens the piglets' passive immunity and shields them from illness. If a sow has more piglets than the number of teats she has, the extra piglets can be placed with another sow with a smaller number of piglets. This is only possible if the sows' piglets are born within a few days of one another. A sow may occasionally refuse to accept her own piglets; this is typically the result of birth hock, which is frequently observed in sows giving birth to their first litter. The piglets may be removed from the sow for a few hours if

Fig. 1.5 Lactating sow

this occurs. They should, if at all feasible, be placed with another sow if she still refuses to accept them. In the event that no other sow is available to raise the rejected piglets, they may be artificially raised on the milk of cows or goats. Additional nutrition options for the piglets include mashed bean porridge with a small amount of sugar added, or goat or cow's milk. However, because the piglets do not always develop and function well in this manner, artificial rearing requires patience and labour-intensive work (Fig. 1.6).

1.7 Nutrition

The importance of vitamins and minerals in swine nutrition was starting to become apparent at the turn of the twentieth century (Forbes 1914). However, because they contained "unidentified growth factors," swine diets were designed with many elements from a practical feeding management perspective (Morrison 1940). Eventually, it was discovered that these "unidentified growth factors" were actually vitamins, minerals, and other vital nutrients. Consequently, fewer components were usually added to swine diets; instead, diets consisting of corn and soybean meal supplemented with vitamins and trace elements were introduced. To see similar trends in understanding of the discrepancies between the requirements of crude protein and individual essential amino acids in pigs as we did for vitamins and minerals. Morrison (1940) evaluated how amino acids were classified as essential and non-essential, leading to the development of the idea of restricting amino acids. Subsequently, Baker et al. (1975) noticed that pigs fed low-crude protein diets supplemented with L-lysine hydrochloride (HCl) showed growth performance comparable to pigs fed high-crude protein diets with the same concentration of total lysine. The idea of the "ideal protein" was developed after studies by Wang and Fuller

Fig. 1.6 Piglets weaning

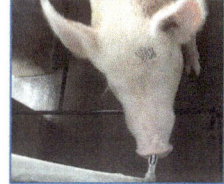

Fig. 1.7 Feeding of greens and concentrate feed with watering

(1989) clarified the advantages of expressing the demand for other amino acids as a ratio relative to lysine. The development of the amino acid ratios allowed for more accurate meal planning, reducing the amount of crude protein consumed while still fulfilling estimations of other amino acid requirements. Crystalline versions of lysine and other amino acids were soon being produced as it was realised how important lysine was as the first limiting amino acid in swine diets. Diets these days frequently call for the inclusion of three to five crystalline amino acids in their formulation. This has resulted in a 40% reduction in nitrogen requirements and a significant drop in nitrogen excretion in swine manure. Standardised ideal digestible amino acid coefficients are now used in diets that were previously designed on a total basis due to variations in the digestibility of crystalline amino acids and intact protein sources (Stein et al. 2007) (Fig. 1.7).

The creation of models to help determine the amino acid requirements and estimate the economic effects of potential changes in diet formulation has benefited from requirement studies. With just around one-third of the total phosphorus produced from plants available to the pig, the creation of the enzyme phytase has significantly reduced the amount of inorganic phosphorus added to swine diets. In 1915, Anderson published one of the earliest accounts of the phytase enzyme in a literary work. However, it wasn't until the early 1970s that noteworthy research on the advantages of phytase was discovered in pigs and poultry, respectively (Jongbloed et al. 1992). Additionally, the shift in emphasis from total requirements to accessible or digestible phosphorus decreased excretion of phosphorus and enabled nutritionists to better meet the pig's development requirements. As a result, phosphorus excreted in swine manure was reduced by 30–40%.

The swine industry has transitioned from using digestible or metabolisable energy to using the net energy system in terms of energy systems (Noblet et al. 1994). These days, diets are designed with the ratio of lysine to calories in mind, and the other amino acids are balanced in relation to lysine (Wang and Fuller 1989). Pork producers have benefited from these advancements in technology and diet formulation strategy because they have increased growth rate, feed efficiency, and carcas leanness, decreased feed costs per pound of gain, and lessened environmental effect by significantly lowering excretion rates. Pork production became more efficient as a result of advances in feed manufacture. Feed efficiency and digestibility were enhanced by pelleting diets and grinding grains to smaller particle sizes. The quantity of nitrogen and phosphorus excreted by pigs has decreased by about 40% as a result of dietary adjustments made over the years, and the carbon footprint of swine production has decreased as well (Tokach and DeRouchey 2013).

1.8 Productive Management

The number of swine farms has decreased as farm sizes have expanded, in line with other mature industries (Giamalva 2014). Despite the socio-economic effects of this shift, larger farms have made it possible to use crucial production techniques that have raised the productivity of the swine herd. Farrow-to-finish farming has given way to multiple-site farming with specialised farms where the breeding herd is kept apart from the expanding pig population. In addition to significantly enhancing health, this shift has made it possible to specialise in work and use numerous health, genetic, reproductive, and nutrition advancements that were previously covered in this chapter. Larger sow farms produce pigs that are more easily handled as a group during the growth season because they have a greater number of pigs who are all the same age, body weight, and health. In addition to the growth benefits of these techniques, group management makes it easier to gather production data, enabling ongoing production enhancement.

The expansion of swine farms has also given production systems the scale required to accommodate research centres. Producers are able to test new technologies and evaluate university discoveries in real-world settings, thanks to the widespread use of commercial research facilities (Tokach et al. 2010). These facilities support the ongoing development that raises productivity even more. In order to achieve the scale or larger farm size, the swine industry has developed a number of

1 Introduction

models; nonetheless, coordinated production is a crucial part of all of these systems. Coordination can take the form of group ownership of sow farms, contract production, or integration to obtain access to other food chain segments. The majority of integration cases, however, involve producer ownership of feed production (cropland) and pigs, which offers a way to boost the value of the grain crop and the manure nutrients from the pig production. Integration is frequently understood as packer ownership of pig production.

1.9 Health Management

Improvements in biosecurity, diagnostic management, animal population control, and the creation of replacement breeding stock devoid of numerous endemic infections have significantly improved the "baseline" health condition of both individual and national herds. The national herd is currently mainly free of endemic illnesses such *Actinobacillus pleuropneumonia*, swine dysentery, progressive atrophic rhinitis, sarcoptic mange, swine brucellosis, trichinosis, and others (USDA 2016). There has been steady progress in getting rid of or controlling other infections including porcine circovirus and mycoplasma pneumonia. The sector has long produced meat free of antibiotic residues, is actively involved in pork quality assurance programmes, and is taking part in the global initiative to lower the use of antibiotics in general (www.pork.org). There are several factors contributing to the national herd's enhanced health (Fig. 1.8).

The capacity of a single sow farm to produce more pigs rose with farm size, opening the door to the possibility of multi-site production. Weaned piglets can be

Fig. 1.8 Affected with swine fever

separated from older animals (sows) in multiple-site production while they still have maternal protection to numerous infections, avoiding these illnesses from impairing performance (Harris 2000). The removal of significant economic diseases from the sow farm is made possible by separating the growing pigs from the sow population. This improves both the reproductive and progeny performance of the pigs that are not exposed to these diseases. Scientific advances in vaccination have played a crucial role in managing illnesses such as porcine circovirus, which is a difficult disease to eradicate from swine farms. Pig health has been significantly improved by increased biosecurity and knowledge of disease transmission, which has limited the entry of diseases through travel, visitors, feed, and breeding stock. There are still issues with health protection in the US swine business. Controlling or eliminating certain diseases becomes more challenging when pig raising sites are concentrated in particular regions of the nation. Some diseases have also found a way to spread due to the mobility and transportation of pigs across the nation (Fig. 1.9).

American diagnostic labs are always coming up with innovative ways to identify disease agents early on and stop them from entering or spreading throughout the swine herd. From gross necropsy to ELISA, PCR, and deep sequencing of bacteria and viruses, these instruments have advanced over time. In order to better safeguard the health of their pigs, farmers have recently benefited from cooperative programmes for disease surveillance and monitoring that have helped them understand how diseases spread across geographical areas.

Fig. 1.9 Biosecurity measures of a farm in front of entry gate

1.10 Conclusion

Through the use of innovative manufacturing technologies, the swine sector keeps developing. It might be feasible to supply the projected demand for animal protein by 2050 if advancements continue at the same pace as they have over the past 30 years. As research efforts progress towards a molecular knowledge of growth and reproduction, fascinating new technologies remain to be uncovered. Remaining connected to the businesses they serve and relevant in basic and applied pig research will be a challenge for colleges. Universities are essential for producing the next generation of swine technologists and for continuing to be a reliable source of independent research that is made available to the general public.

References

Anderson RJ (1915) The hydrolysis of phytin by the enzyme phytase contained in wheat bran. J Biol Chem 20:475

Baas TJ (1996) ISU Swine Enterprise Records Program. Swine Research Report, 1996. Paper 31. http://lib.dr.iastate.edu/swinereports_1996/31

Baker DH, Katz RS, Easter RA (1975) Lysine requirement of growing pigs at two levels of dietary protein. J Anim Sci 40(5):851–856. https://doi.org/10.2527/jas1975.405851x

Bortolozzo FP, Menegat MB, Mellagi APG, Bernardi ML, Wentz I (2015) New artificial insemination technologies for swine. Reprod Domest Anim 50(2):80–84. https://doi.org/10.1111/rda.12544

Dawkins MS (2017) Animal welfare with and without consciousness. J Zool 301:1–10

Forbes EB (1914) 1914 Mineral metabolism experiments with swine. Proc Am Soc Anim Nutr 1:4–6. https://doi.org/10.2134/jas1914.191414x

Giamalva J (2014) Pork and swine industry and trade summary. US International Trade Commission. Control number 2014002. https://www.usitc.gov/publications/332/pork_and_swine_summary_its_11.pdf

Harris DL (2000) Multiple site production. Iowa State Univ. Press, Ames. https://doi.org/10.1002/9780470376935

Jongbloed AW, Mroz Z, Kemme PA (1992) The effect of supplementary *Aspergillus niger* phytase in diets for pigs on concentration and apparent digestibility of dry matter, total phosphorus, and phytic acid in different sections of the alimentary tract. J Anim Sci 70:1159–1168. https://doi.org/10.2527/1992.7041159x

Ketchem R, Rix M (2013) Using production data to make decisions. http://benchmark.farms.com/2013_Using_Production_Data.html

Kim JS, Yang X, Pangeni D, Baidoo SK (2015) Relationship between backfat thickness of sows during late gestation and reproductive efficiency at different parities. Acta Agric Scand Sect A Anim Sci 65(1):1–8

Krogh U, Oksbjerg N, Purup S, Ramaekers P, Theil PK (2016) Colostrum and milk production in multiparous sows fed supplementary arginine during gestation and lactation. J Anim Sci 94(Suppl 3):22–25

Miao ZH, Glatz PC, Ru YJ (2004) Review of production, husbandry and sustainability of free-range pig production systems. Asian Australas J Anim Sci 17:1615–1634

Morrison FB (1940) Feeds and feeding, 20th edn. The Morrison Publishing Company, Ithaca, NY

Noblet J, Fortune H, Shi XS, Dubois S (1994) Prediction of net energy value of feeds for growing pigs. J Anim Sci 72:344–354. https://doi.org/10.2527/1994.722344x

Stein HH, Sève B, Fuller MF, Moughan PJ, de Lange CFM (2007) Invited review: amino acid bioavailability and digestibility in pig feed ingredients: terminology and application. J Anim Sci 85:172–180. https://doi.org/10.2527/jas.2005-742

Thomas LL, Goodband RD, Tokach MD, Dritz SS, Woodworth JC, DeRouchey JM (2018) Partitioning components of maternal growth to determine efficiency of feed use in gestating sows. J Anim Sci 96:43134326

Tokach M, DeRouchey J (2013) Feeding swine and poultry low protein diets with feed-use amino acids and the effect on the environment. Feedstuffs, 23 April. http://mydigimag.rrd.com/article/Feed-Use_Amino_Acids_Beneficial/1040799/108615/article.html

Tokach MD, Dritz SS, Goodband RD, DeRouchey JM, Nelssen JL (2010) Where has all the research gone? In: 2010 Al Leman swine conference proceedings. http://nationalhogfarmer.com/genetics-reproduction/where-has-research-gone-1115/

USDA (2016) Swine 2012. Part II: reference of swine health and health management in the United States, 2012. National Animal Health Monitoring Systems, Fort Collins, CO. https://www.aphis.usda.gov/animal_health/nahms/swine/downloads/swine2012/Swine2012_dr_PartII.pdf

Wang TC, Fuller MF (1989) The optimum dietary amino acid pattern for growing pigs. Br J Nutr 62:77–89. https://doi.org/10.1079/BJN19890009

Breeds of Pigs of India and its Productivity

2

N. Rajanna, J. Saikiran, and J. Shashank

Abstract

Pigs are quite important in India, especially for the lower socioeconomic sections of society. Total number of pigs in India during the year 2019 was 9.06 million (BAHS, 2023). Due to some innate characteristics like as high fecundity, improved feed conversion efficiency, early maturity, and short generation interval, pigs have a greater potential to provide farmers with a faster economical returns than the other livestock species. In India pork production is constrained, constituting merely 9% of the nation's animal protein sources. As per ICAR-NBGAR at present 13 registered indigenous pig breeds are present in India. Apart from the registered breeds, some population of indigenous pigs are also found in India.

Keywords

Pig breeds · India · Age at first farrowing · Hot carcass weight

India has world's largest livestock population and the livestock sector contributes a significant quarter to the agricultural gross domestic product. Within the spectrum of livestock species, pigs hold a crucial position, particularly among the weaker sections of society. Pigs exhibit substantial potential to provide rapid economic returns to farmers, owing to inherent traits such as high fecundity, efficient feed conversion, early maturity, and a short generation interval when compared to other livestock species. Additionally, pig farming demands relatively small investments in buildings and equipment. This sector offers enormous potential to secure both nutritional and economic well-being for the needy segments of society

N. Rajanna (✉) · J. Saikiran · J. Shashank
ICAR-Krishi Vigyan Kendra, Mamnoor, PVNRTVU, Warangal, TG, India

© The Author(s), under exclusive license to Springer Nature Singapore Pte Ltd. 2024
T. Rana, B. Soto-Blanco (eds.), *Good Practices and Principles in Pig Farming*, Livestock Diseases and Management,
https://doi.org/10.1007/978-981-97-4665-1_2

As per 2023 BAHS, exotic/crossbred pigs' population was 1.90 million and indigenous pigs' population was 7.16 million contributing total pig population to 9.06 million. The percentage change from 2012 to 2019 was −12.03. India produces very little pork; it makes up only 9% of the country's animal protein sources. The focal point of production lies predominantly in the northeastern region, where it is largely carried out by backyard and informal sector producers. The market for processed pork products in India is limited, with a majority of the demand being met through imports. While there are a few local companies engaged in the production of processed items like sausages and bacon, the quantities produced are restricted, and the industry remains relatively small.

2.1   Strength in Swine Production

High potential for profit and revenue. Piggeries can be set up in relatively compact spaces. Feed costs are considerably lower compared to other meat production expenses. The demand for pork meat has seen substantial growth, driven by elevated prices and limited availability of red meat substitutes. The turnaround production time is faster than red meat production, making it a preferred meat choice.

2.2   Opportunities

Increasing demand; Access to venture capital; Enhancement of value and export opportunities; A means for poverty alleviation; Opportunities for self-employment; An industry with significant growth potential.

2.3   Weakness

Cultural taboos; Limited progress in breed upgrading; Lack of availability of concentrated feed; Fragile supply chain and marketing facilities; Insufficient meat processing infrastructure; Higher labor intensity compared to other segments of the meat industry; Lack of a National Traceability Program.

Most of the native pigs were categorized as Desi, local, or unremarkable. There were few published references that classified them into populations like the Ghoongroo pigs of West Bengal, the Ankamali pigs of peninsular India, the Gahuri pigs of Assam, and the Desi pigs of North India, as reported by Bhat et al. (1981, 2010). However, recent research by Boro et al. (2016a) has led to the characterization and documentation of new populations. Some of these have been identified as distinct and have been officially recognized as new breeds of indigenous pigs by ICAR-NBAGR.

2.3.1 Registered Breeds of Indigenous Pigs

At present 13 breeds are registered by ICAR-NBAGR. Ghoongroo, Niang Megha, Agonda Goan, Tenyi Vo, Nicobari, Doom, Zovawk, Gurrah, Purnea, Mali, Manipur black, Wak Chambi, and Banda are the 13 registered breeds of indigenous pigs in India.

2.3.1.1 Ghoongroo

With an estimated number of 15,000, these Ghoongroo breed pigs may be found in several districts of West Bengal, such as Darjeeling, Jalpaiguri, Cochbehar, North Dinajpur, and South Dinajpur (Behl et al. 2002). The Ghoongroo pigs exhibit distinctive characteristics, being black-colored, with concave snouts and pendulous ears. In males, the height at withers ranges from 66 to 95 cm (84 cm on average), while in females, it is between 59 and 92 cm (78 cm on average). The litter size at farrowing is typically 12, with a range of 8–18.

According to a study conducted by Sahoo et al. (2012), these pigs typically experience their first heat at around 190.38 ± 4.38 days of age. However, Gokuldas et al. (2015) suggested a slightly higher age of puberty, noting it as 7.8 ± 0.41 months. The gestation period for Ghoongroo pigs is around 113.8 ± 0.16 days, and the age at first farrowing is reported to be 13.1 ± 0.65 months. The farrowing interval is approximately 213.53 ± 0.4 days.

The reported litter size at birth is 0.96 ± 0.02, which then increases to 8.7 ± 0.25 at weaning, according to studies conducted by Sahoo et al. (2012) and Gokuldas et al. (2015). In terms of hot carcass weight, males typically range from 85 kg (63–90 kg), while females range from 82 kg (61–90 kg), as stated in the Breed Descriptor of 2013. Musculus longissimus thoracis et lumborum fatty acid profiling reveals that the proportion of saturated and unsaturated fatty acids ranges from 32.17% to 41.19% and 58.98% to 68.15%, respectively. Furthermore, the presence of Omega-3, -6, and crucial fatty acids can be observed at levels ranging from 0.88% to 1.73%, 19.95% to 27.23%, and 14.52% to 23.47%, respectively, as stated by Thomas et al. (2016). In terms of subcutaneous fat analysis, the proportion of saturated and unsaturated fatty acids is reported to be 34.99% and 61.72%, respectively, according to Naskar et al. (2014).

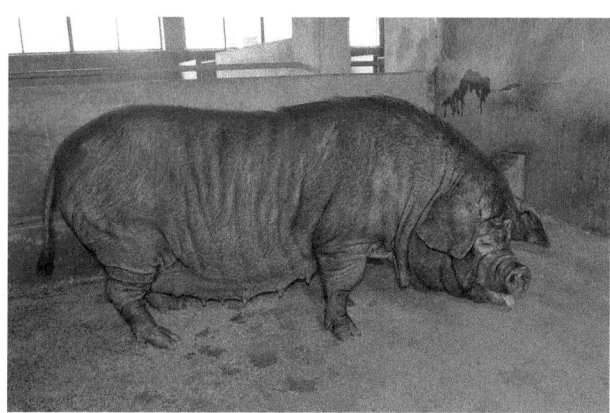

2.3.1.2 Niang-Megha

The Niang-Megha breed of pigs is predominantly found in Meghalaya, characterized by a mostly black coat, elongated snouts, short erect ears, and pot bellies. In males, the height at withers is recorded at 51.4 cm (with a range of 48–54 cm), while in females, it is 48.2 cm (with a range of 45–52 cm). The population of Niang-Megha pigs is estimated as approximately 3.9 lakhs (Behl et al. 2002).

The reported age at which these pigs experience their first heat is 210.5 ± 2.42 days. The age at first farrowing is 355.25 ± 2.25 days, and the farrowing interval is approximately 207.05 ± 8.16 days. The gestation period for Niang-Megha pigs is noted to be 111.85 ± 0.14 days.

At birth, the weight is recorded at 0.485 ± 0.23 kg, while at weaning it is 4.97 ± 0.21 kg. The litter size at birth is 6.34 ± 0.26, which decreases to 5.63 ± 0.42 at weaning (Sahoo et al. 2012; Khargharia et al. 2014; Gokuldas et al. 2015).

The hot carcass weight for males ranges from 30 to 58 kg, with an average of 40.60 kg, while for females, it ranges from 32 to 60 kg, with an average of 44.10 kg (Behl et al. 2002). Fatty acid profiling of musculus longissimus thoracis et lumborum indicates that the range of saturated fatty acids is 32.75–38.91%, and for unsaturated fatty acids, it is 54.69–69.28% (Thomas et al. 2018).

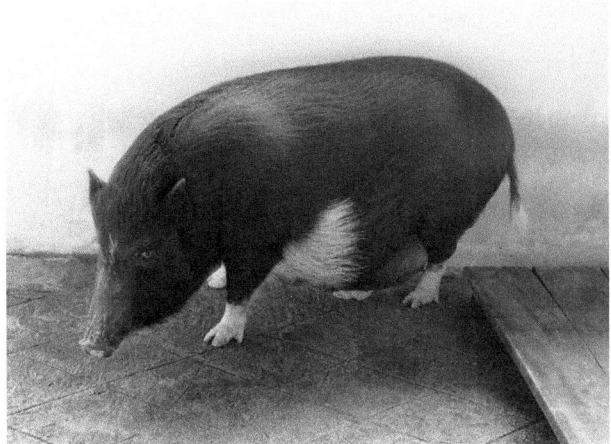

2.3.1.3 Agonda-Goan

These swine are predominantly found in Tiswadi, Bardez, and Peenem in the North Goa district, as well as Marmugoa, Salcete, Canacona, and Quepem taluks in the South Goa district, with an estimated population of 37,556. The Agonda-Goan pigs exhibit a predominantly black coloration, with short and erect ears, short snouts, and pot-bellied physiques. In males, the height at withers is recorded at 55.87 cm, ranging from 46 to 64.5 cm, while in females, it is 60.77 cm, with a range of 48.2–74.8 cm.

The reported age at which Agonda-Goan pigs first give birth is 334.75 ± 27.32 days, and their farrowing interval is 185.35 ± 4.11 days. The litter size at the time of farrowing is recorded as 7.45 ± 0.51, while at weaning, it is 5.9 ± 0.55. According to the Breed Descriptor (2015), the carcass weight for males is reported to be 48.30 ± 1.48 kg, while for females, it is 48.60 ± 2.47 kg.

2.3.1.4 Tenyi Vo

The Tenyi Vo breed of pigs is primarily found in Nagaland, displaying a predominantly black coloration with distinctive white patches on their bellies. These pigs are distinguished by their long, straight noses, erect ears, concave top lines, pot-bellied bodies, and white patches surrounding the nostrils. The height at withers in males is measured at 30.48 cm, with a range of 28 to 36 cm, and in females, it is measured at 25.4 cm, with a range of 23–30 cm. The population of this group is estimated to be between 60,000 and 70,000, spread out in Kohima, Peren, Phek, and Dimapur districts of Nagaland (Breed Descriptor 2016).

The age at first farrowing for Tenyi Vo pigs is reported as 298.35 ± 2.03 days, with a corresponding farrowing interval of 149.99 ± 2.36 days (Chusi et al. 2015). The litter size at birth is noted as 5.8 ± 2.3, and at weaning, it is 4.2 ± 1.9 (Borkotoky et al. 2014). Additionally, the litter weight at birth is reported as 1.91 ± 1.92 kg, increasing to 25.60 ± 2.54 kg at weaning (Chusi et al. 2015). The male pigs have a recorded hot carcass weight of 38 kg (35–40 kg), while the female pigs have a recorded weight of 36 kg (35–40 kg) (Breed Descriptor 2016).

Male animals of the Tenyi Vo breed have been observed to mature at a very early age, with the ability to impregnate sows even before reaching the age of 6 months (Kumaresan et al. 2007; Karunakaran et al. 2009).

2.3.1.5 Nicobari

Nicobari pigs are mainly found in the Nicobar district of Andaman and Nicobar, with an estimated population of 35,000. The Nicobari breed of pigs predominantly displays black or brown coloring, with a small percentage exhibiting creamy white, reddish brown, or a mix of black and brown. These pigs are characterized by short ears, medium-sized pot bellies, short to medium snout with a slight concave conformity and concave top lines.

In males, the height at withers is measured at 67.31 cm, ranging from 51 to 86 cm, while in females, it is 61.24 cm, with a range of 50–82 cm. The age at first farrowing for Nicobari pigs is reported as 319.19 days, with a farrowing interval ranging from 260 to 450 days. The litter size at farrowing is noted as 7.19, with a range of 4–12, and at weaning, it is 5.42, with a range of 3–8. The male pigs have a recorded hot carcass weight of 52.6 kg (24–96 kg) while the female pigs have a recorded weight of 58.53 kg (44–81 kg) (Breed Descriptor 2016).

2.3.1.6 Doom

Primarily these pigs are found in Agomani, Gauripur, Golakganj blocks, and Bilasipara sub-division in Dhubri district, as well as in select areas of Bongaigaon and Kokrajhar districts of Assam, with an estimated population of approximately 3000 (Breed Descriptor 2016). The Doom breed pigs are characterized by their black color, erect ears, concave snout, and straight top lines.

In males, the height at withers ranges from 60.4 to 66.2 cm, while in females, it is between 65 and 71 cm. The age at first farrowing for Doom breed pigs is reported as 368.0 ± 1.54 days, and the farrowing interval is 213.53 ± 0.4 days. The litter size at farrowing is noted as 6.25 ± 0.24, and at weaning, it is 5.03 ± 0.21 (Khargharia et al. 2014). The male hot carcass weight is documented at 32.04 ± 1.89 kg, while females weigh in at 35.10 ± 1.2 kg according to the Breed Descriptor (2016).

2.3.1.7 Zovawk

Zovawk pigs, indigenous to Mizoram, exhibit a distinctive black coloration and are characterized by their small size and bulging bellies. These pigs have short, erect ears and a short, concave snout. In males, the height at withers is measured at 49.15 cm, with a range of 45–54 cm, while in females, it is 49.04 cm, with a range of 47–50 cm. Zovawk pigs are distributed mostly across Mamit, Kolasib, Aizawl, Lunglei, Sahia, and Champhai districts of Mizoram, boasting an estimated population of around 40,000.

The age at first farrowing for Zovawk pigs is reported as 314.9 days, with a farrowing interval ranging from 249 to 350 days (Breed Descriptor 2017). Kalita et al. (2018) specifically reported that the average age at first fertile service was 323.75 ± 9.90 days, the average age at first furrowing was 437.75 ± 9.41 days, the gestation period was 113.63 ± 0.53 days, and the service periods were 113.13 ± 7.81 days. The litter size at farrowing is noted as 7.40 ± 0.40, and at weaning, it is 5.2 ± 0.66. Carcass weight is reported at 28.29 kg (23–33 kg) in male animals and 34.2 kg (25–43 kg) in female animals (Breed Descriptor 2017).

Male animals of the Zovawk breed also exhibit early maturation, being capable of impregnating sows even before reaching 6 months of age (Kumaresan et al. 2006).

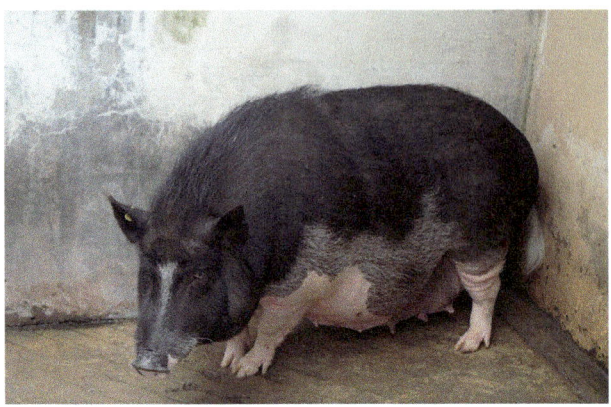

2.3.1.8 Gurrah

These pigs are primarily located in the Bareilly division and adjoining areas of the Lucknow division in Uttar Pradesh, as reported by the National Bureau of Animal Genetic Resources in 2018 (NBAGR 2018). Boro et al. (2016b) have also provided a description of the phenotypic characteristics of Bareilly Desi pigs found in the Bareilly region. These pigs display a mature body weight of 53.1 ± 0.40 kg for males and 53.5 ± 0.40 kg for females.

Gurrah pigs are characterized by their black color, have legs that are white below the hock joints. Medium-sized with an angular body and a flat belly. The head is elongated with a triangular face, featuring a long and straight snout. A thick line of hair is present on the neck from the head to the shoulders. The ears are short, vertically erect and leaf-shaped. Adult Gurrah pigs typically weigh about 46 kg in males and 48 kg in females. The litter size at farrowing is 6.85, and at weaning, it is 5.65.

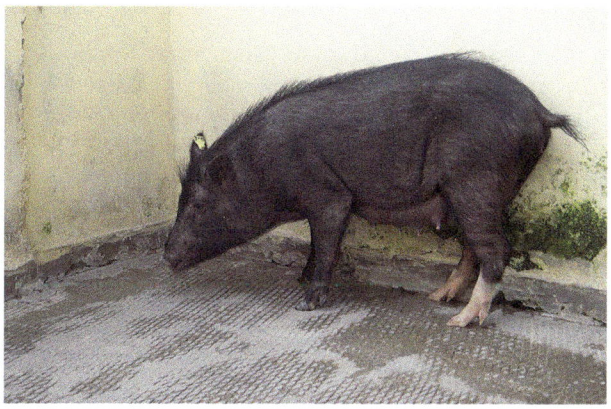

2.3.1.9 Purnea

Purnea, a medium-sized pig with a distinctive black color. Generally these are found in the Purnea and Katihar districts of Bihar and in the adjoining areas of Sahibganj district in Jharkhand. On their lower limbs some individuals may exhibit white spots. These pigs possess a compact body and a pot belly, with a thick line of bristles running along the topline from the neck to the shoulders, imparting a wild appearance. They are characterized by a round face, slightly concave snout short conical and erect ears. Mature animals have thick skin with neck folds.

Adult Purnea pigs typically weigh between 41 and 50 kg, and their litter size at birth varies from 4 to 6. Despite their compact and pot-bellied physique, some individuals may display white spots on their lower limbs. These pigs are renowned for their fierce temperament, with an estimated population ranging from 100,000 to 120,000.

2.3.1.10 Mali

These pigs are predominantly found in the Dhalai and North districts of Tripura. They exhibit a black color with white patches, many individuals have a star on their forehead and medium-sized with pot belly. These animals feature short, a concave snout and erected ears lying perpendicular to the body axis. The average adult body weight is around 50 kg. The litter size ranges between 8–10 at birth and 7–8 at weaning. The initial farrowing age is estimated to be 281.4 ± 1.6 days, while the farrowing gap is roughly 178.5 ± 0.9 days, according to studies conducted by Dandapat et al. (2010) and Naskar et al. (2013). The population size is estimated to be around 45,000 to 50,000.

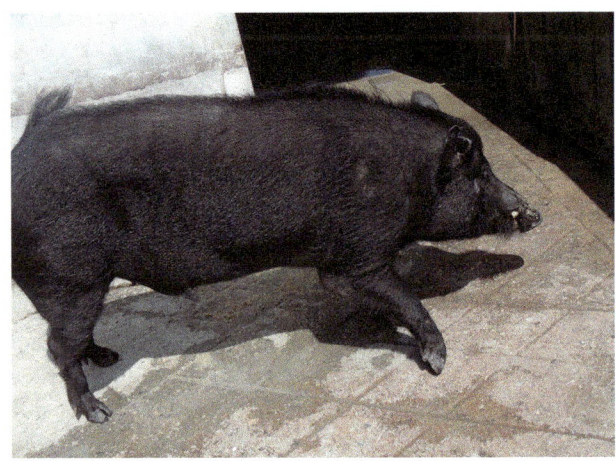

2.3.1.11 Manipur Desi

These pigs are primarily found in Manipur and are occasionally referred to as Burmese pigs by the local population. They are characterized by drooping ears which are large-sized. Medium-sized pigs possessing an adult body weight ranging from 120 to 150 kg, and the litter size is typically about 8–12, according to Naskar et al. (2013).

The Manipuri Black pig originates from Manipur state and, as its name implies, has a black coat. It is of medium size, with a flat stomach and short legs. The head is short and somewhat concave, with short ears and a snout of short to medium length. White spots can sometimes be seen on the extremities like the legs and snout. Most of the pigs have black fur with some having a dark grey coloration. There is minimal bristle growth, and trimming is not a common practice. Adult body weight average in males is about 96.0 kg, while females average approximately 93.0 kg. The average litter size is 8.27 (with a range of 6–11) at birth and 6.02 (with a range of 5–9) at weaning.

2.3.1.12 Wak Chambil

Wak Chambil pigs are known for their small size and distinctive round, medium pendulous belly. These pigs can be found in different areas of Meghalaya, such as North Garo Hills, East Garo Hills, South Garo Hills, West Garo Hills, and Southwest Garo Hills. They have a small head and eyes, small erect ears, and a short, pointed snout. In addition, Wak Chambil pigs have thick, long hair on their eyebrows, forehead, and neck. Their limbs are short, with small hooves that touch the ground partially. The bristles on these pigs are short and dense, covering their entire body.

In terms of weight, adult males average around 32.0 kg, while females average approximately 29.0 kg. The average litter size is 5.8 (with a range of 4–11) at birth and 4.52 (with a range of 3–8) at weaning. Notably, the pork from this breed is known for its unique flavor and taste, making it especially significant during special religious and ceremonial occasions.

2.3.1.13 Banda

The Banda pig, indigenous to Jharkhand, exhibits a black coloration and features short, erect ears. These animals have medium to short bristles on their necks and a long, concave snout. Pot-bellied in nature, Banda pigs are known for their small litter size. Adult males average around 28.0 kg, while females average approximately 27.0 kg. The average litter size is 4.5 (with a range of 4–7) at birth and 4.25 (with a range of 4–6) at weaning.

2.3.2 Other Population of Indigenous Pigs of India

Apart from the above described registered breeds, some other population of indigenous pigs have been described.

2.3.2.1 Ankamali

They take their name from a location near Ernakulum, and Ankamali pigs are predominantly found in Kerala, with distribution extending to Karnataka and Tamil Nadu. These particular pigs exhibit a combination of black and rusty grey coloring, along with patches of white. They have a small and compact body structure, with a slight bulge at the belly. The face is elongated, featuring a tapering snout, and there is a noticeable bulge at the joint of the jaws. The back has a slight concave shape, and by the time they reach 8 months of age, they typically weigh around 37 kg. The reported weights at birth, 8, and 32 weeks of age were 0.86 ± 0.03, 8.03 ± 0.16, and 46.37 ± 0.93 kg, respectively (Gupta et al. 2007; Naskar et al. 2013).

The Dome pigs, on the other hand, are distributed in the North and South districts of Tripura. Their color varies from black to grey, featuring a long, thick crest of bristles along the dorsal line from the neck to the trunk. Adult body weight is approximately 50 kg, and the litter size ranges from 8–10 at birth to 5–7 at weaning (Naskar et al. 2013; Das et al. 2018).

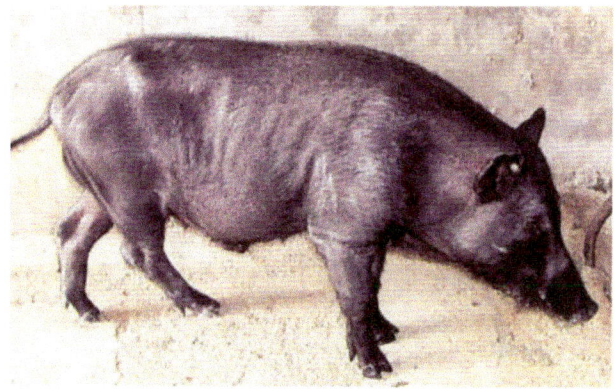

2.3.2.2 Golla

Named after the Golla community in the Ganjam district of Odisha, these pigs are traditionally raised by this community. They are of medium-sized, with an adult weight ranging from 50 to 80 kg. The litter size is about 8–10, and there are two farrowing events per year, as indicated by Naskar et al. (2013).

2.3.2.3 Lepchamoun

Lepchamoun pigs are raised in Sikkim and are of medium size, featuring drooping ears. Adult animals typically weigh between 80 and 120 kg. The reported litter size at birth is around 7–10, according to Naskar et al. (2013). Nath et al. (2013) reported that the litter size at birth and at weaning for these local pigs in Sikkim, which was reported as 4.3 ± 0.45 and 2.79 ± 0.24, respectively. The age at first farrowing is 365.39 ± 7.96 days, and the farrowing interval is 196.27 ± 8.37 days.

References

Basic Animal Husbandry Statistics (2023) available at https://dahd.nic.in/sites/default/filess/BAHS2023.pdf
Behl R, Vij PK, Niranjan SK, Bhel J, Vijh RK (2002) Indigenous pig genetic resources of India: Distribution, types and their characteristics. Indian J Anim Sci 90(2): 127–133
Bhat PN, Batt PP, Khan BU, Goswami OB, Singh B (1981) Animal genetic resources of India. National Dairy Research Institute, Karnal, pp 75–83
Bhat PN, Mohan NH, Deo S (2010) Pig production. Studium Press (India) Pvt Ltd., Darya Ganj, New Delhi, p 21
Borkotoky D, Perumal P, Singh RK (2014) Morphometric attributes of Naga local pigs. Vet Res Int 2:8–11
Boro P, Patel BHM, Naha BC, Sahoo NR, Gaur GK, Dutt T, Singh M, Madkar A (2016a) Productive and reproductive performances of desi pigs: a review. Agric Rev 37(3):228–233
Boro P, Patel BHM, Sahoo NR, Naha BC, Madkar A, Gaur GK, Singh M, Dutt T, Verma MR, Upadhyay D, Singh AK (2016b) Phenotypic attributes of Bareilly *desi* pig. Int J Adv Biol Res 6:390–393
Breed Descriptor (2015) New breeds of indigenous livestock and poultry: AgondaGoan pigs. Indian J Anim Sci 85:546–548
Breed Descriptor (2016) New breeds of indigenous livestock: TenyiVo, Nicobari and doom pigs. Indian J Anim Sci 86:1221–1225
Breed Descriptor (2017) New breeds of indigenous livestock: Zovawk pigs. Indian J Anim Sci 87:1427–1428

Chusi Z, Savino N, Dhali A, Perumal P (2015) Phenotypic morphometric parameters of indigenous pig of Nagaland. Indian J Anim Sci 85:1334–1337

Dandapat A, Dev CKB, Debbarma C, Das MK (2010) Phenotypic characterization of Mali pig in Tripura, India. Livest Res Rural Dev 22:article 83

Das S, Naha BC, Saini BL (2018) Adopted way of pig rearingpractices in Tripura. J Entomol Zool Stud 6(4):1673–1678

Gokuldas PP, Tamuli MK, Mohan NH, Barman K, Sahoo NR (2015) A comparative analysis of reproductive performance of different pig breeds under intensive management system in sub-tropical climate. Indian J Anim Sci 85:1042–1045

Gupta N, Ahlawat SPS, Behl R, Behl J, Vijh RK, Singh G and Gupta SC (2007) Pig genetic resources of India: Ankamali—A Pig breed of Indian Peninsula. National Bureau of Animal Genetic Resources, Karnal. Monogram 44/2007.

Kalita G, Sarma K, Rahman S, Talukdar D, Ahmed FA (2018) Morphometric and reproductive attributes of local pig of Mizoram. Int J Livest Res 8:173–177

Karunakaran M, Mondal M, Rajarajan K, Karmakar HD, Bhat BP, Das J, Bora B, Baruah KK, Rajkhowa C (2009) Early puberty of local Naga boar of India: assessment through epididymal spermogram and *in vivo* pregnancy. Anim Reprod Sci 111:112–119

Khargharia G, Zaman G, Laskar S, Das B, Aziz A, Roychoudhary R, Roy T (2014) Phenotypic characterization and phenotypic studies of NiangMegha and doom pigs of north eastern India. Asian Acad Res J Multidiscip 27:667–676

Kumaresan A, Hussain J, Ahmed SK, Pathak KA, Das A, Bujarbaruah KM (2006) Growth performance of Hempshire, Large White Yorkshire and Mizo local pigs under Mizoram field conditions. Indian J Anim Sci 76:148–150

Kumaresan A, Bujarbaruah KM, Pathak KA, Chhetri B, Das SK, Das A, Ahmed SK (2007) Performance of pigs reared under traditional tribal low input production system and chemical composition of non conventional tropical plants used as pig feed. Livest Sci 107:294–298

Naskar S, Niranjan SK, Bani S (2013) Utilization of pig genetic resources in India. In: Pundir RK, Niranjan SK, Behl R (eds) Sustainable utilization of indigenous animal genetic resources. NBAGR, Karnal, pp 120–125

Naskar S, Mandal GP, Borah S, Vashi Y, Thomas R, Dhara SK (2014) Evaluation of fatty acid profile in subcutaneous adipose tissue of indigenous and crossbred pigs. Indian J Anim Sci 84:88–90

Nath BG, Pathak PK, Ngachan SV, Tripathy AK, Mohanty AK (2013) Characterization of small holder pig production system: productive and reproductive performance of local and crossbred pigs in Sikkim Himalayan region. Trop Anim Health Prod 45:1513–1518

NBAGR (2018) New breeds registered. National Bureau of Animal Genetic Resources. Available at: http://www.nbagr.res.in/ registered breeds.html.

Sahoo NR, Das A, Naskar S, Banik S, Tamuli MK (2012) Niang-Megha: the nature's gift for food and fiber, ICAR- National Research Center on Pigs, Rani, Guwahati. Pp 1–30

Thomas R, Banik S, Barman K, Mohan NH, Sharma DK (2016) Profiles of colour, minerals, amino acids and fatty acids of musculus longissimus thoracis et lumborum of Ghungroo pigs. Indian J Anim Sci 86:1176–1180

Thomas R, Banik S, Barman K, Mohan NH, Sharma DK (2018) Selected meat quality parameters and nutritional profiles of M. musculus longissimus thoracis et lumborum of Niang- Megha pigs. Indian J Anim Sci 88:955–960

Indigenous Pigs Breeds of Nepal

Subir Singh

Abstract

The chapter explores the significant role of pig farming in Nepal, where the population stands at approximately 1.5 million, contributing substantially to annual pork production, totaling 29,493 metric tons. Despite this contribution, pork consumption remains relatively low, with per capita consumption at a mere 1 kg, representing 5.5% of the total meat consumption. Indigenous pig breeds, particularly Chwanche, Hurrah, and Bampudke, dominate the landscape, constituting around 81% of the total pig population. These indigenous breeds exhibit favorable attributes such as hardiness, adaptability to local conditions, and suitability for low-input systems, making them vital for subsistence farmers.

Moreover, the chapter delves into the origin of Nepalese pigs, tracing their lineage to the Eurasian wild boar. It discusses the genetic diversity of domestic pigs and emphasizes the importance of preserving indigenous breeds for their unique characteristics and sociocultural significance. Detailed descriptions of indigenous pig breeds such as Chwanche, Hurrah, and Bampudke are provided, highlighting their phenotypic characteristics, growth and reproductive performance, and socioeconomic importance. Despite their lower meat production compared to exotics, indigenous breeds offer distinct advantages such as adaptability to local conditions and cultural significance.

The chapter concludes by discussing future prospects and challenges in Nepalese pig farming, including the need for conservation efforts to protect endangered breeds like Bampudke. It underscores the importance of integrating modern technologies like artificial insemination to enhance the productivity of

---

S. Singh (✉)
Department of Veterinary Medicine and Public Health, Faculty of Animal Science, Veterinary Science and Fisheries, Agriculture and Forestry University, Rampur, Chitwan, Nepal
e-mail: ssingh@afu.edu.np

© The Author(s), under exclusive license to Springer Nature Singapore Pte Ltd. 2024
T. Rana, B. Soto-Blanco (eds.), *Good Practices and Principles in Pig Farming*, Livestock Diseases and Management,
https://doi.org/10.1007/978-981-97-4665-1_3

traditional breeds while preserving their genetic diversity. Overall, the chapter provides valuable insights into the dynamics of pig farming in Nepal, highlighting its significance for food security, livelihoods, and cultural heritage.

Keywords

Nepal · Indigenous pig · Breeds · Chwanche · Hurrah · Bampudke

3.1 Introduction

Nepal has a population of about 1.5 million pigs and annual pork production in the country is 29,493 MT contributing about 5.34% of total national meat production (MoALD 2021). The current per capita meat consumption is 18.1 kg, with buffalo meat accounting for 58% of the total meat consumed. Pork, on the other hand, now occupies the fourth position with a meager 1 kg per capita contribution (5.5% of the overall meat consumption). About 10% of the household (9.9%) keep pigs in the country (Singh and Chapagain 1998). This figure is higher in eastern hilly region where ethnic people and socially disadvantaged are populated. Since pig is the rapid converter of feed to meat protein, pig is considered as an important commodity for achieving food security for the rapidly growing human population (ADS 2015). Pigs are primarily maintained for pork and provide animal protein to human beings and their role in progressive agriculture is by providing manure for maintaining soil fertility and to meet sociocultural beliefs (Shakya 2008).

Pigs are small-sized livestock contributing to poor and subsistence farmers who can't afford to rear large-sized livestock which demand high price per head and cost of maintenance. The subsistence farmers rear mainly the indigenous pigs in their backyard using mainly the kitchen leftovers. The indigenous pig breeds in the country cover around 81% (Chwanche 53%, Hurrah 23%, Bampudke <1%) and the exotic origin represents about 19% of total pig population (Rasali et al. 1998) suggesting the more contribution comes from indigenous breeds.

According to recent data, native pigs make up 58% of the overall pig population, with exotic or improved breeds accounting for the remaining 42% (Kayastha 2006). With the technology intervention and the change in attitude towards pig, the production of pig meat is increased drastically. Production of pig meat of Nepal increased from 4300 thousand MT in 1970 to 29,493 thousand MT in 2020 which is growing each year. The demand of pork meat production is rising annually by 10% in Nepal exhibiting high scope in pork industry. One of the main reasons for the rise in pork demand is removal of cultural barriers that prevented people from consuming the meat. The demographic distribution of indigenous pig breeds of Nepal shows the highest population of pig is found in mid hills (53%) followed by Terai (36%) and high hills (11%) (MoALD 2020). Two pig breeds, the Hurrah from the east and the Bampudke from the west of Nepal (Gorkhali et al. 2021).

In the traditional subsistence agriculture system, pigs are raised mostly by indigenous people using scavenging techniques close to the communities. The indigenous pig breeds are Hurrah, Chwanche, and Bampudke (Neopane and Kadel 2008). Under this arrangement, it has been discovered that several ethnic communities are engaged in the pig industry. Although the system's intake and output are both modest, they both significantly contribute to the food and nutrition security of households as well as the satisfaction of religious requirements. In some communities, pigs are even used as gift particularly during the marriage ceremony. Pig's meat (pork) used to be consumed by certain communities in the country in the past.

The commercial pig production system, which has emerged recently and is expanding quickly, rears pigs in confinement with commercial feed supplementation, primarily using exotic breeds and their many hybrids. Hampshire, Yorkshire, Landrace, and Duroc are the popular alien breeds that were brought in for industrial use. Even young Brahmin men are discovered to be engaged in commercial piggery, proving that tribes and religion have no bearing on this industrial method. With the idea of factory farm production, farmers are discovered to be keeping everything from a small number of breeding sows to as many as 200 breeding sows and over 1000 fatteners. In addition to these systems, it has been discovered that fish farming and pig production are interwoven in the nation (Gurung et al. 2014).

3.2 Origin of Nepalese Indigenous Pigs

The Eurasian wild boar (*Sus scrofa*) is the ancestor of the domestic pig. It has been established beyond doubt that domestication of wild pig subspecies in Europe and Asia happened on its own. It is thought that the ancestral forms diverged some 500,000 years ago, which is a long time before they were domesticated some 9000 years ago. Asian pigs were brought to Europe in the eighteenth and early nineteenth centuries, according to historical accounts. A significant contribution to the field of pig genetics, the study will affect how genetic diversity in this livestock species is maintained and used (Giuffra et al. 2000).

There have been at least 16 distinct subspecies of wild boar proposed, and they are found in Northwest Africa and throughout Eurasia (Ruvinsky and Rothschild 1998). It's possible that local populations of wild boars periodically domesticated pigs (Bokonyi 1974). Whether modern farmed pigs that exhibit noticeable physical differences from their wild ancestor have a single or several origins is still up for debate. Darwin distinguished between two primary domestic pig species: the Asian (*Sus indicus*) and European (*Sus scrofa*). The adoption of Asian pigs to enhance European pig breeds in the eighteenth and early nineteenth centuries is well-documented (Darwin 1896; Jones 1998). However, it is unclear how much genetic material Asian pigs have contributed to various European pig breeds. Nepal is fortunate to have a wild pig population with grey to brown bodies, rough hair, and later-vanishing brown stripes down the sides of the piglets' bodies. These are said to be the progenitor of domestic pigs in Nepal. Research is necessary to establish a solid foundation for demonstrating the origin of domestic pigs in Nepal.

3.3 Importance of Indigenous Pig Breeds of Nepal

There are three native pig breeds in the nation that have been classified and identified to varying degrees. Hurrah, Chwanche, and Bampudke are their names (Shrestha 1995; Neopane 2004).

The Chwanche, who live in the hills, make up a significant share of the population among Indigenous breeds (Shrestha 1996). A total of 58% of the pigs are Chwanche, which live in the highlands (Rasali et al. 1998; Funk et al. 2007). According to Rasali et al. (1998) and Funk et al. (2007), hurrah pigs make up 23% of the overall pig population and are found in the terai region. Bampudke, also known as *Sano Badel* and *Pigmy hog* (Epstein 1977), which are found both in wild and domestic form are now in the stage of risk category of existence, i.e., about to be extinct (Neopane 2006). Government organization has taken replacement strategy to improve the productivity of national pig population, indigenous pigs' population is reduced to 58% of the total pig population while the remaining 42% are exotic or improved breeds (Kayastha 2006).

3.3.1 Sociocultural Significance

The two main categories of the nation's pig production system are newly established commercial piggeries and subsistence (traditional) production in scavenging management (Douge et al. 1989).

3.3.2 Positive Attributes

In general, indigenous breeds are low producing than the exotics in terms of meat production. However, they have several other positive attributes such as hardiness, adaptability to local harsh conditions and can produce in low input system. It is observed that these positive attributes are not duly recognized rather exotics breeds are unnecessarily given preference over them.

3.3.3 Negative Attributes

On the down side, pigs—specially those raised under scavenging management systems—are frequently held accountable for environmental contamination. The town has serious issues as a result of these animals polluting water sources and leaving garbage scattered across the neighborhood. Likewise, the loudness and unpleasant smell that pig rearing in a human population produces might cause social cohesion to be disturbed. The fact that these animals can aid in the spread of many zoonotic diseases is the most crucial factor. But, with careful management, the chance of illness spreading and conflict arising from unpleasant smells and loudness can be reduced (Table 3.1).

3.4 Features of Indigenous Nepalese Pig Breeds

3.4.1 Chwanche Pig (*Sus domesticus*)

Chwanche pigs are found in low and middle hills regions of Nepal and is a good scavenger animal. The variety is also known as "Pundi" in the eastern hills (SARP 1996). They are hardy, resistant to diseases and well adapt to their local habitat (DLS 2016). Chwanche is a poor man's pig raised by under privileged people of hilly region in scavenging situation (Gorkhali et al. 2021).

Key phenotypic characteristics: Chwanche pigs are striking black in color and have long, straight noses. Their ears are small and erect, and their barrel resemble the drooping type. Females usually have 8–12 teats that support their young, while average weight of an adult's ranges between 25 kg and 40 kg, with an average of 35 kg. Their tails are long and straight, contributing to their small body size and short stature. Their length, measured from the tip of their head to the base of their tail, reaches 76 cm, and their heart girth reaches 86 cm. These pigs exhibit semi-wild tendencies in their natural habitat. The average body length, height at withers, height at hipbone, and heart girth of this breed is 75.9 ± 2.1 cm, 55.9 ± 1.84 cm,

Table 3.1 Showing features of indigenous pig breeds of Nepal

S.N.	Breeds	Scientific name	Home tract	Population status	Characterization	Positive attributes
1.	Chwanche	*Sus domesticus*	Across the hills	Population declining but not yet at risk	Phenotypic + chromosomal + DNA	Suitable for hills, disease Resistant, hardy, suitable for Backyard rearing
2.	Hurrah	*Sus domesticus*	Across the Terai	Population declining But not yet at risk	Phenotypic + chromosomal + DNA	Suitable for Terai, strong body, Hardy, suitable for backyard Rearing
3.	Bampudke	*Porcula salvania*	Few terai districts near Chure Hills (Nawalparasi, Chitwan, Dang, Kailali etc.)	Risk (about to be Extinct)	Phenotypic + chromosomal + DNA	Smallest hog breed, both wild And domestic, quality of meat

Fig. 3.1 Chwanche pig (Source: Gorkhali et al. 2021)

54.6 ± 1.51 cm, and 86.3 ± 3.37 cm respectively. The average adult weight of a pig is 35 (25–40) kg (Gorkhali et al. 2021) (Fig. 3.1).

Growth performance and reproductive performance: Chwanche gave birth to her first progeny at an age of 10.7 months, with an average weight of 0.7 ± 0.06 kg. The weight of 1 year is approximately 33.2 ± 0.08 kg. The Chwanche pigs are the medium-sized pigs amongst the indigenous breeds. The first mating age, first farrowing age and farrowing interval of Chwanche pigs are 7.3 ± 0.85, 10.7 ± 0.80, and 7.4 ± 0.6 months, respectively. The average litter size of this breed is 7.33 ± 1.28 (Gorkhali et al. 2021).

3.4.2 Hurrah Pig (*Sus domesticus*)

The Terai region of Nepal, which is tropical and subtropical, is home to wild pigs, which are primarily raised for meat. They are resilient and best suited for the wealthy residents of the Terai region who are in a state of scavenging (Fig. 3.2).

Key phenotypic characteristics: Body color is completely grey-black or rusty brown, skin is boasts rough, ears are small and erect. Pigs has straight snout complements with the straight and brown brittles hair adorning its neck and back. The pigs bear dropping barrel type, it stands tall on long, sturdy legs, its tail straight and long. Females have 8–12 teats that support their offspring. Average adult weight is 45 kg, ranging from 40 kg to 55 kg, top length is 79 cm, heart girth is 88 cm. Pigs display semi-wild behavior and remain calm in their natural environment. The average body length, withers height, hipbone height, and heart girth of this breed is 79.5 ± 1.75 cm, 61.1 ± 1.2 cm, 62.1 ± 1.07 cm, and 87.7 ± 2.18 cm, respectively. The average adult weight of Hurrah pig is 45 (40–55) kg (Gorkhali et al. 2021).

Growth performance and reproductive performance: Hurrah gives birth to the first generation of piglets with an average weight of 0.8 ± 1.2 kg at 14 months of

Fig. 3.2 Hurrah pig (Source: Gorkhali et al. 2021)

age. The weight of 1 year is approximately 41.6 ± 0.09 kg. The Hurrah pigs are the largest pig among local species. Hurrah pig's age at first mating, age at first calving, and calving interval by month are 10.8 ± 0.99, 14.0 ± 0.96, and 5.57 ± 0.57 months, respectively. The average litter size of Hurrah pig is 7.04 ± 1.26 (Gorkhali et al. 2021).

3.4.3 Bampudke Pig (*Porcula salvania*)

The Bampudke pig, also known as the Pigmy hog, is one of the smallest pigs in the world and one of the hardest breeds. The species' historical range followed the narrow alluvial tract south of the Himalayan foothills, connecting central Assam in the east with south-eastern Uttarakhand in the west, and the Indian states of Uttar Pradesh (UP), Bihar, West Bengal, and Assam with Nepal and Bhutan's international borders (Oliver 1980; Oliver and Deb Roy 1993).

Key phenotypic characteristics: Pigs usually appear in shades of red and brown, sometimes appearing in black. Their length (from head to base of tail) is approximately 45 cm, and their heart girth is 52 cm. Females have 8–12 teats that support their offspring. The average weight of an adult pig is 20 kg, with a range of 18–25 kg. These indigenous pigs have a high fertility rate and exhibit attractive characteristics such as litter size and farrowing interval. Bampudke pigs are found in both domesticated and wild forms, demonstrating their ability to adapt to their environment. The average body length, withers height, hipbone height, and heart girth of this breed is 45.7 ± 3.30 cm, 39.6 ± 2.71 cm, 34.5 ± 2.94 cm, and 51.8 ± 2.42 cm, respectively (Gorkhali et al. 2021) (Fig. 3.3).

Growth performance and reproductive performance: Bampudke pigs give birth at the age of 11.5 months, with an average weight of 0.6 ± 0.08 kg piglets. The average adult weight of a pig is 20 (18–25) kg. The Bampudke pigs are the smallest-sized pigs amongst the indigenous breeds. The first breeding age, first farrowing age and farrowing interval of Bampudke pigs were determined as 6.2 ± 0.12, 11.5 ± 0.23,

Fig. 3.3 Bampudke pig (Source: Gorkhali et al. 2021)

and 4.6 ± 0.28 months, respectively. The average litter size of this species is 4.7 ± 0.27 (Gorkhali et al. 2021).

3.5 Future Prospects

The country's average pig production is increasing day by day. Despite the high demand for pork, pig farming is ignored by most communities. Domestic pigs make up 80% of the pig population. Native breeds are popular due to their quality. Domestic pigs can be raised successfully with lower investment than exotic breeds and are well received in the clay soils of Nepal. Pork is popular throughout the country, although their meat yield is lower than exotic breeds, local breeds are popular among consumers. Native breeds have other advantages, such as cold tolerance and adaptability to harsh local conditions. While many species have proven their ability to thrive in systems with low external input, different species are unable to cope. For example, traditional pigs (Chwanche and Hurrah) were collected and raised from poor tribes in the highlands and Terai. They play an important role in rural life in the Terai and hilly regions of Nepal. The religious, cultural and economic values of some mountain tribes are additional features for their protection. In order for normal birth to continue, there must be some special training and birth rules.

Domestic pig breeding has made a great contribution to meat production. In addition, these animals provide fertilizer that is used to increase the fertility of the soil. Farmers now raise large numbers of pigs to expand their business and sustain their lifestyle. Despite the rapid growth of the pork industry, farmers deal with the mix and match of indigenous pigs. Bampudke pig is the smallest animal in Nepal and is on the verge of extinction. There is an urgent need to develop and implement plans and programs for the protection of domestic pigs in the region. The production of our pigs is always low and it is necessary to make a profit by competing with some exotic breeds. Artificial insemination is the best tool for preserving traditional breeds and making them fertile by mixing them with exotic breeds.

References

Agriculture Development Strategies (ADS) (2015) Government of Nepal, Ministry of Agriculture development. 363

Bokonyi S (1974) History of domestic mammals in central and Eastern Europe. Akademiai Kiado, Budapest, p 597

Darwin F (1896) The letters of Charles Darwin. Nature 55(1418):196–196

DLS (2016) Annual report

Douge K, Kurusawa Y, Tanaka K, Nishida T, Pradhan SM, Raj Bhandary HB (1989) A study on karyotype of the Indian wild pig and domestic native pig in Nepal. In: Morphological and genetical studies on the native domestic animals and their wild forms in Nepal. Part II. Published by Faculty of Agriculture, The University of Tokyo, Tokyo, pp 95–101

Epstein H (1977) Domestic animals of Nepal. Holmes & Meier Publishers, New York, London, pp 72–75

Funk SM, Verma SK, Larson G, Prasad K, Singh L, Narayan G, Fa EJ (2007) The pygmy hog is a unique genus: 19th century taxonomists got it right first-time round. Mol Phylogenet Evol 45:427–436

Giuffra EJMH, Kijas JMH, Amarger V, Carlborg Ö, Jeon JT, Andersson L (2000) The origin of the domestic pig: independent domestication and subsequent introgression. Genetics 154(4):1785–1791. https://www.researchgate.net/publication/326070385_Pygmy_hog_Porcula_salvania_Hodgson_1847

Gorkhali NA, Sapkota S, Bhattarai N, Pokhrel BR, Bhandari S (2021) Indigenous livestock breeds of Nepal: a reference book. Published by National Animal Breeding & Genetics Research Centre, NASRI, NARC, Lalitpur

Gurung TB, Gurung TB, Shrestha BS, Shrestha NP, Bates R, Neupane D, Achhami K (2014) Pig and pork industry in Nepal. In: In proceedings of the 1st National Workshop on pig and pork industry in Nepal, 10–11 December 2013, Kathmandu Nepal

Jones GF (1998) Genetic aspects of domestication, common breeds and their origin. In: Ruvinsky A, Rothschild MF (eds) The genetics of the pig. CAB, London, pp 17–50

Kayastha KP (2006) A scenario on pig production in Nepal: present situation challenges in treatment and elimination of Taeniasis/Cysticerosis in Nepal, pp 47–54

MoALD (2020) Statistical information on Nepalese agriculture, 2013/2014. Government of Nepal, Ministry of Agriculture Development, Agricultural Statistics Division, Singhadurbar Kathmandu

MoALD (2021) Statistical information on Nepalese agriculture, 2013/2014. Government of Nepal, Ministry of Agriculture Development, Agricultural Statistics Division, Singhadurbar Kathmandu

Neopane SP (2004) Native Animal Genetic Resources of Nepal: status of their Conservation and Utilization. In: Proceedings of IV National Conference on Science and Technologies. Nepal Science and Technology (NAST), Kathmandu, pp 74–88

Neopane SP (2006) Characterization of indigenous animal genetic resources of Nepal. In: Proceedings of the 6th National Workshop on livestock and fisheries research, Nepal. Agricultural Research Council, Kathmandu, pp 1–11

Neopane SP, Kadel R (2008) Indigenous pigs of Nepal. National Animal Science Research Institute & Nepal Agriculture Research Council, Khumaltar, Kathmandu

Oliver WLR (1980) The pigmy hog: the biology and conservation of the pigmy hog *Sus* (Porcula) *salvanius* and the hispid hare *Caprolagus hispidus*. The Jersey Wildlife Preservation Trust special scientific report

Oliver WLR, Deb Roy S (1993) The pigmy hog (*Sus salvanius*). In: IUCN/SSC pigs and peccaries specialist group and IUCN/SSC hippo specialist group. Pigs, Peccaries and Hippos Status Survey and Action Plan

Rasali DP, Pradhan SM, Dhaubdel TS (1998) In: Shrestha JNB (ed) Pig genetic resources. Proceedings of the first National Workshop on animal genetic resources conservation and genetic improvement of domestic animals in Nepal. Nepal Agricultural Research Council, Khumaltar, Lalitpur, pp 33–39

Ruvinsky A, Rothschild MF (1998) Systematics and evolution of the pig. In: Ruvinsky A, Rothschild MF (eds) The genetics of the pig. CABI, Wallingford, pp 1–16

Shakya P (2008) Social inclusion as a pre-condition of development in Nepal. Leisa Magazine, Lalitpur

Shrestha NP (1995) Animal genetic resources of Nepal and their conservation. In: Proceedings of the third global conference on conservation of domestic animal genetic resources. RBI Canada, Toronto, pp 113–119

Shrestha NP (1996) Animal genetic diversity of Nepal. Proceedings of First National Workshop on Livestock and Fisheries Research in Nepal. Nepal Agricultural Research Council Khumaltar, Lalitpur, pp 55–61

Singh SB, Chapagain DP (1998) Livestock sector in the agriculture perspective plan. In: Proceedings of the first national workshop on animal genetic resources conservation and genetic improvement of domestic animals in Nepal-April 11–13, 1994 (Ed. J.N.B. Shrestha). Nepal Agricultural Research Council, Khumaltar, pp 117–128

Swine and Avian Research Program (SARP) (1996) Annual report. SARP, NARC

Body Condition Scoring of Pigs

Vallabhaneni Srikanth and Cherryl Dimphna Miranda

Abstract

In the intricate and dynamic world of swine production, the well-being and health of pigs play a vital role in the overall success of the farm. A scoring system for physical condition must be utilized to constantly monitor the productivity of the swine herd. Body condition scoring (BCS) system is one such important tool in swine production management which is used to assess the nutritional status, performance, stress levels, and welfare status of pigs. It can be measured either by visual, subjective or objective methods. In practice most commonly, the BCS score in pigs ranges between 1 and 5. Excessive emaciation is indicated by BCS 1, extreme obesity by BCS 5 and the sow's reproductive performance is at its best around BCS 3. Body condition scoring gives a clear picture of the adequacy of the feed and the efficiency of the feed distribution system, which can be extremely helpful in managing sows in any scenario. At weaning, mid-gestation, farrowing and service, sows should have their conditions assessed. Thus, it implies that in addition to extending swine longevity, maintaining optimal BCS can enhance swine reproductive success and thereby overall farm profits.

Keywords

BCS · Body condition · Health · Management · Performance · Production · Sow

V. Srikanth
Livestock Production & Management Section, ICAR-Indian Veterinary Research Institute (ICAR-IVRI), Izatnagar, Bareilly, Uttar Pradesh, India

C. D. Miranda (✉)
Department of Livestock Production Management, Veterinary College, Athani, Karnataka, India

Veterinary Animal and Fisheries Sciences University, Bidar, Athani, Karnataka, India

© The Author(s), under exclusive license to Springer Nature Singapore Pte Ltd. 2024
T. Rana, B. Soto-Blanco (eds.), *Good Practices and Principles in Pig Farming*, Livestock Diseases and Management,
https://doi.org/10.1007/978-981-97-4665-1_4

4.1 Introduction

The swine industry has witnessed remarkable advancements in production efficiency, genetics, and management practices over the years. As the demand for high-quality pork products continues to rise, there is an increasing emphasis on optimizing the health and welfare of pigs throughout their production lifecycle. Among the various metrics used for assessing the well-being of pigs, body condition score (BCS) has emerged as a valuable tool for evaluating nutritional status and overall health.

In the case of sows, determining body composition through direct dissection is impractical, especially in field conditions. Indirect methods, such as deuterium dilution or measuring backfat and live weight (Knudson et al. 1985), provide alternatives. Bio-electrical impedance, as utilized in growing pigs (Swantek et al. 1992), is another option. However, these indirect approaches, while relatively precise, pose challenges when applied on a large scale in farm settings. Consequently, as a substitute, the assessment of animal body condition using scoring techniques has been recommended in various species including sheep, cattle (Evans 1978; Edmonson et al. 1989), horses (Carrel and Huntington 1988), and pigs (Patience and Thacker 1989). Body scoring methods provide information on conformational changes at various body sites through text descriptions, illustrations, and photos.

BCS is a subjective yet widely adopted method that allows producers and researchers to estimate the body fat reserves of pigs. It involves palpating specific anatomical regions to assign a score that reflects the amount of subcutaneous fat present. While BCS has been extensively utilized in other livestock species, its application in pigs has gained prominence in recent years due to its potential to inform nutritional management decisions, reproductive performance assessments, and disease detection. As we delve into the intricacies of body condition scoring for pigs, it is evident that a nuanced understanding of this tool holds the key to unlocking new avenues for improved swine health, welfare, and productivity.

4.2 What Is BCS?

Body condition scoring (BCS) in pigs is a systematic and standardized method for quantifying the degree of fatness and muscularity in an individual pig through the visual and tactile assessment of specific anatomical landmarks such as backfat thickness, muscle development, and bony prominences. This subjective evaluation is assigned a numerical score, typically on a predetermined scale, providing a practical and noninvasive means of gauging the nutritional status and overall health of pigs in agricultural and research settings.

The BCS technique combines three key elements in swine husbandry:

- Ensuring proper welfare.
- Implementing effective husbandry practices.
- Achieving optimal performance.

4.3 Importance of Body Condition Scoring of Pigs

Body condition scoring (BCS) of pigs is crucial for several reasons and it plays a significant role in pig management and welfare. Here are some key reasons highlighting the importance of body condition scoring in pigs:

1. Enhanced Nutritional Management.
2. Sustaining Optimal Reproductive Performance.
3. Routine Health Monitoring.
4. Strategic Resource Allocation and Management.
5. Monitoring Economic Viability in Pig Farming.
6. Adaptation to Varied Environmental Conditions.
7. Mitigation of Obesity and Overweight-Related Concerns.
8. Advancement in Animal Welfare.
9. Optimization of Market Weight.
10. Augmented Farm Productivity and Profitability.

Regular monitoring of the swine herd's productivity necessitates the use of a physical condition scoring system. Accurately evaluating body condition empowers farm managers to provide optimal nutrition to animals, minimizing feed wastage, and effectively controlling feed costs. The implementation of body condition grading is instrumental in enhancing the reproductive efficiency of the sow herd. Sows must maintain good bodily condition throughout their lives to ensure both contented living and maximal productivity for the swineherd. A well-run operation prioritizes feeding sows to sustain appropriate body condition scores (BCS) across all production stages, recognizing the impact of good body condition on both short- and long-term output potential.

A sow's physical condition score serves as a predictive indicator of both her current productivity and future prospects. Underweight sows may face health and quality of life challenges, potentially leading to delayed estrus and inadequate support for healthy fetal development or pregnancy. Conversely, overweight sows may encounter difficulties in caring for their piglets, including issues such as leg problems, challenges during farrowing, smaller litters, reduced feed intake during lactation, and lighter weaning weights. Significant weight loss during nursing is associated with problems like smaller litter sizes, lower conception rates, and delayed ovulation post-weaning. Research also indicates that over- or under-conditioned sows, particularly during lactation when minerals are mobilized from bones, have a higher likelihood of developing bone and spine issues. Condition scores offer valuable insights for adjusting daily feeding rates during gestation, ensuring each sow is appropriately prepared for farrowing and re-breeding. This proactive care not only enhances sow longevity but also improves reproductive efficiency.

The body condition score of an animal may fluctuate over the course of the year, influenced by factors such as the environment and available feed. To maintain good health, it is advisable to assess the body condition score periodically depending on the production and life cycle of the animals. Mastering the assessment of pigs' physical condition through body condition scoring is an indispensable tool for increasing the overall productivity and profitability of a swine herd on a farm.

Factors affecting the body condition scoring: When assessing the body condition of an animal, it is crucial to consider three factors (Neary and Yager 2002):

1. **Gut fill**, including stage of pregnancy: Gut fill (involving feed and water intake) and pregnancy stage can impact the body condition score, with animals in full or late gestation potentially being mistakenly scored higher, while those fasting may be scored lower than their true condition.
2. **Amount of hair or wool**: The thickness of hide, hair, or wool on some animals may make scoring challenging without manual palpation.
3. **Amount of muscle**: Heavily muscled animals may appear round, potentially causing confusion with fat deposition, while light-muscled animals might be mistakenly perceived as thin. Assessing muscle expression is best done through the center of the round (or hindquarter), where fat has less influence. Animals with bulge and flair typically exhibit greater musculature, while angular animals tend to be lighter muscled.

Additional elements encompass nutrition, encompassing both the amount and quality of food, weather conditions and surroundings, genetic factors, overall health and specific diseases, the stage of production such as age or gestation compared to lactation, and the proficiency of caregivers, among other considerations.

Body condition scoring in pigs: When evaluating the body condition of a sow to determine its body fat levels, the primary areas of attention should include the face, shoulders, pelvis, and topline. A scoring system, such as the three-point, five-point, or six-point scale, can be utilized to assign scores based on this assessment which are listed in the following tables (Table 4.1, 4.2, 4.3 and 4.4)

The five-point scoring method is most commonly used for evaluating pigs. In body condition scoring, it is necessary to palpate various parts of the sow's body to assess the fat covering. The scoring system employs a scale from 1 (emaciated) to 5 (obese), incorporating both visual evaluation and tactile examination. Relying solely on visual assessment is inadequate; physically handling the pig is crucial for an accurate condition assessment. The scoring process emphasizes the ease of palpating specific body checkpoints, depicted below (Fig. 4.1):

Table 4.1 Three-point scale (Renggaman et al. 2015)

Body condition score	Description
0	Animal with a good body condition
1	Animal with moderate body condition
2	Animal with a poor body condition (lean animals)

Table 4.2 Five-point scale (Coffey et al. 1999a, b)

BCS score	Status of animal
1	Emaciated/very thin
2	Thin
3	Ideal/good
4	Fatty
5	Very fat/obese

Table 4.3 Six-point scale type 1 (Whay 2008)

BCS score	Condition
1	Very thin
2	Thin
3	Insufficient
4	Tight
5	Good
6	Obese

Table 4.4 Six-point scale type 2 (Hutu and Onan 2020)

BCS score	Description
0	Emaciated/very thin
1	Poor/thin
2	Moderate
3	Ideal/good
4	Fat
5	Grossly fat/obese

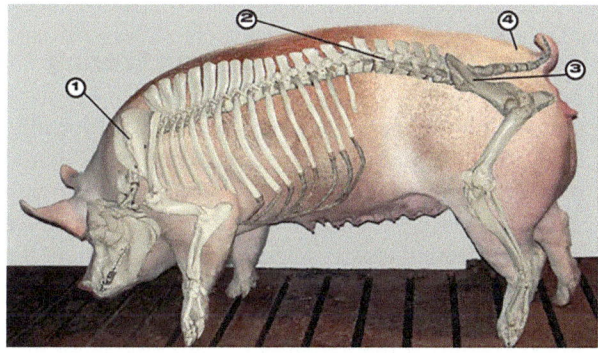
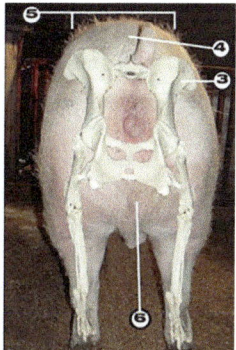

Fig. 4.1 Skeletal points that should be palpated or visually identified for body condition scoring. 1. Shoulder blades. 2. Spine. 3. Hip bones. 4. Tail head. 5. Top shape. 6. Between legs (seam of hams). (Adapted from: https://www.nationalhogfarmer.com/hog-health/sow-body-condition-scoring-guidelines)

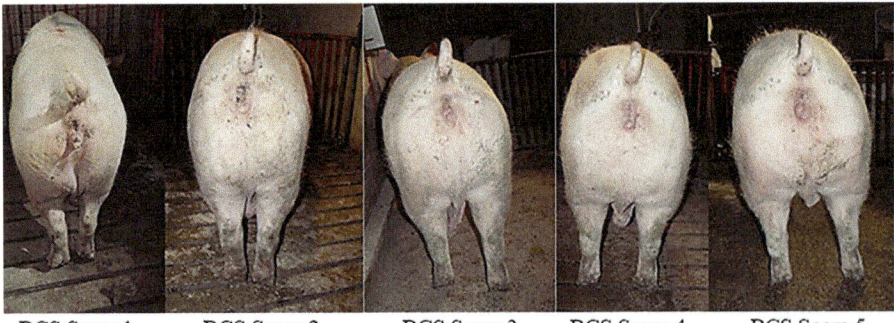

BCS Score 1   BCS Score 2   BCS Score 3   BCS Score 4   BCS Score 5

Fig. 4.2 Pig body condition scoring system employed in the general context. BCS Score 1. BCS Score 2. BCS Score 3. BCS Score 4. BCS Score 5. (Photo credit: Dr. Ken Stalder and the National Hog Farmer magazine)

Table 4.5 Body shape of pigs and BCS

BCS number	Condition of pig	Body shape of pig
5	Overfat	Bulbous
4	Fat	Tending to bulge
3.5	Good condition	Tube shaped
3	Normal	Tube shaped
2.5	Somewhat thin	Tube shaped but flat (slab) sides
2	Thin	Ribs and spine can be felt
1	Emaciated	Bone structure apparent (ribs and backbone)

4.4 Visual Appraisal of Pigs During Body Condition Scoring

The level of body fat rises progressively on the scale from one to five. Typically, pigs accumulate fat starting from the front and moving towards the back. Initial weight gain occurs in the jowl, shoulder, and topline. Subsequent weight increase leads to the filling out of the ribs, flank, belly, and hips, resulting in a rounder shape and diminished visible muscle tone. During the assessment of body condition, the evaluator examines and feels for key bone structures.

As swine experience weight loss in the pelvic region, certain bone structures, such as the tail head and the ilium or hook of the pelvic girdle, become more pronounced. Further weight loss causes a reduction of flesh from the hindquarters, along with a diminishing presence around the flanks, ribs, and scapular bones (shoulder). In cases of extreme weight loss, fat becomes noticeably absent from the jowl, revealing distinct jawlines and facial bones (Fig. 4.2).

The visibility of ribs, spinal processes, and hipbones is only notable in pigs with a body condition score of one or those considered emaciated, which is inappropriate for any stage of production. Conversely, an increase in body condition score corresponds to an augmentation of flesh around the pelvis, creating a fuller and rounder appearance.

Body shape of pigs vary from BCS 1–5 which can be depicted in the following Table 4.5 (Jackson and Cockcroft 2007):

4.5 Methods of Evaluating Body Condition Score (BCS) in Pigs

They can be categorized into subjective and objective approaches.

4.5.1 Subjective Methods

This involves a manual assessment, wherein trained evaluators visually examine and palpate specific points on the pig, such as the pelvic bone, ribs, vertebrae, and tail head. This method relies on careful observation and palpation of fat tissue between the skin and bones, with a recommended scale ranging from 1 to 5 points. Table 4.6 provides a basic overview of body condition grading, which ranges from 1 to 5.

Advantages of the subjective method include its simplicity, cost-effectiveness, and applicability on a large commercial scale. However, there are disadvantages, such as the potential for inaccuracy due to the evaluator's skill, overlooking changes over time within the same herd, and challenges when dealing with multiple pig breeds with varying conformation.

4.5.2 Objective Methods

On the other hand, objective methods rely on precise measurements and comprises the following techniques

4.5.2.1 Sow Body Condition Caliper

In 2015, Knauer and Baitinger introduced the sow body condition caliper (Fig. 4.3). This method focuses on assessing the sow's loose fat and muscle. The principle underlying the sow body condition caliper is that it measures the sow's back's angularity. An increase in weight, fat, and muscle causes a sow's back to become more angular. Sows with a body condition score of 1 exhibit a more angular back, whereas those with a score of 5 display a flatter and wider back. Ideal condition with caliper assessment is between 12 and 15 units range.

The caliper is applied to three specific locations on the sow's back: behind the shoulder, middle of the back, and at the last rib. The last rib (Fig. 4.3a, b) is preferred for measurement due to its consistent anatomical location, ensuring accurate assessment. This approach is both rapid and precise in determining sow body condition, effectively minimizing known variations in body condition scores among individual sows. The evaluator positions themselves behind the sow, carefully locating the last rib through palpation. Subsequently, the two arms of the caliper are aligned with the last rib, gently resting on the loin edge of the sow, and a reading is taken (Knauer and Baitinger 2015).

Table 4.6 General assessment of the pigs from the BCS observation

BCS score	1	2	3	4	5
General body condition	Poor, bony, Skeletal	Adequate, slim, Lean	Ideal, normal, Fit	Unsatisfactory, Plump, round	Poor, Overweight, Round
Pin bones or hip bones	Easily visible, very prominent	Pin bones obvious but some slight cover.	Good cover, pin bones only felt with firm pressure	Well padded, impossible to feel	Round hips and heavily covered hipbones making it impossible to feel
Tail head	Deep cavity around tail setting	Cover around tail setting	No cavity around tail	Root of tail set deep in surrounding fat	Cannot be felt
Abdomen/ loin	Very narrow. Sharp edges on transverse spinal process. Flank very hollow	Loin narrow. Only slight cover to edge of transverse spinal process. Flank rather hollow	Edge of transverse spinal process only felt with firm pressure. Flank full	Impossible to feel bones. Flank full and rounded	Rounded and further deposition of fat impossible
Backbone	Vertebrae prominent and sharp throughout length of backbone	Visible with some cover	Good cover, vertebrae only felt with firm pressure	Impossible to feel vertebrae	Midline appears as slight hollow between rolls of fat
Ribs	Individual ribs very prominent	Rib cage is apparent but less prominent than above	Rib cage not visible. Very difficult to feel any ribs	Ribs impossible to feel	Thick fat cover
Rear view of the pigs	1	2	3	4	5

4.5.2.2 Flank-to-Flank Measurement

It involves the use of a cloth tape, is an easy and accurate way to estimate BCS based on body weight. The flank-to-flank measurement is made by going across the top of the sow from the bottom of the left flank to the bottom of the right flank

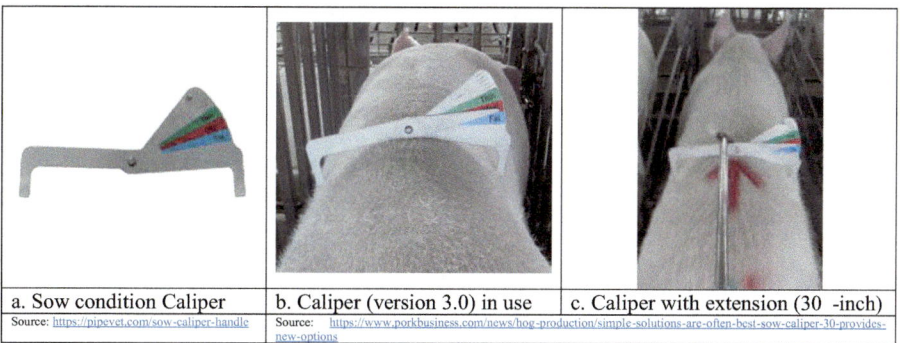

a. Sow condition Caliper	b. Caliper (version 3.0) in use	c. Caliper with extension (30 -inch)
Source: https://pipevet.com/sow-caliper-handle	Source: https://www.porkbusiness.com/news/hog-production/simple-solutions-are-often-best-sow-caliper-30-provides-new-options	

Fig. 4.3 Sow body condition caliper. (**a**) Sow condition caliper. (Source: https://pipevet.com/sow-caliper-handle). (**b**). Caliper (version 3.0) in use. (**c**). Caliper with extension (30-inch). (Source: https://www.porkbusiness.com/news/hog-production/simple-solutions-are-often-best-sow-caliper-30-provides-new-options)

Fig. 4.4 The flank-to-flank measurement in sow (Iwasawa et al. 2004)

(Fig. 4.4). Table 4.7 displays the relation between measured inches and body weights (Iwasawa et al. 2004), while Table 4.8 shows the association with BCS (Young and Aherne 2005).

$$\text{Formula}: \text{Weight}(\text{lb}) = 26.85 \times \text{flank} - \text{to} - \text{flank}(\text{in.}) - 627.93$$

(Iwasawa et al. 2004)

4.5.2.3 The Renco Lean Meter

It is a durable ultrasound-based device, measures backfat thickness with a probe and digital display, requiring an experienced person for accurate assessment. Lean meter comprises a digital display and a probe (Fig. 4.5). The backfat is measured carefully by positioning the probe 7–9 cm from the midline of the last rib. The

Table 4.7 Sow weight (in pounds and kilograms) based on flank-to-flank length (inches) +/− 30 lb

Flank-to-flank measurement		Sow weight	
In inches	In centimeters	In pounds	In kilograms
34	86.36	285.0	129.3
35	88.90	311.8	141.4
36	91.44	338.7	153.6
37	93.98	365.5	165.8
38	96.52	392.4	178.0
39	99.06	419.2	190.2
40	101.60	446.1	202.3
41	104.14	472.9	214.5
42	106.68	499.8	226.7
43	109.22	526.6	238.9
44	111.76	553.5	251.0
45	114.30	580.3	263.2
46	116.84	607.2	275.4
47	119.38	634.0	287.6
48	121.92	660.9	299.8
49	124.46	687.7	311.9
50	127.00	714.6	324.1

Source: Iwasawa et al. 2004

Table 4.8 Correlation of flank-to-flank measurement with BCS

Flank to flank (cm)	Weight category	Estimated weight (kg)	BCS
83–90	Very light	115–150	1
91–97	Light	150–180	2
98–104	Medium	180–215	3
105–112	Heavy	215–250	4
105–112	Very heavy	250–300	5

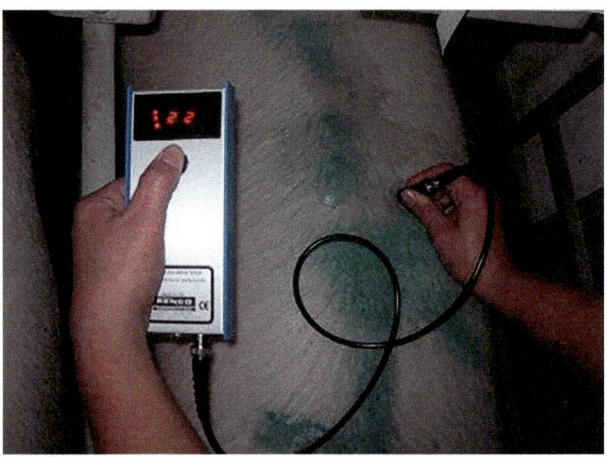

Fig. 4.5 Renco lean meter (https://secrepro.com/en/product/renco-lean-meter/)

Table 4.9 Backfat thickness and BCS relationship

Score	Last rib backfat depth (in.)	Condition	Body shape
1	< 0.6	Emaciated	Hips, spine prominent to the eye
2	0.6–0.7	Thin	Hips, spine easily felt without pressure
3	0.7–0.8	Ideal	Hips, spine felt only with firm pressure
4	0.8–0.9	Fat	Hips, spine cannot be felt
5	> 0.9	Overfat	Hips, spine heavily covered

digital display shows the measurement in millimeters or inches. The association between backfat thickness and BCS (Coffey et al. 1999a, b) is given in the following Table 4.9.

4.5.2.4 Ultrasonography

Backfat ultrasonography using ultrasound technology to measure backfat thickness at specific locations. While B-mode ultrasound can provide precise measurements of the loin eye area or loin depth, both A-mode and B-mode ultrasound technology can provide reliable measurements of backfat. This method provides a more accurate and objective measurement compared to manual palpation.

4.5.2.5 Image Analysis

Digital imaging captures images of pigs and using software for image analysis can be a precise and objective way to assess body condition. It allows for measurements of backfat thickness and muscle area.

4.5.2.6 Computed Tomography (CT) Scans

Whole-body CT scans provide detailed information about fat distribution and muscle mass throughout the entire body. This method is more sophisticated and may be used in research settings.

4.5.2.7 Bioelectrical Impedance Analysis (BIA)

BIA devices measure the resistance of electrical flow through body tissues, providing an indirect measure of body composition. While not commonly used in field conditions, it can offer insights into body fat and lean tissue percentages.

4.5.2.8 Weight-Based Scoring

Weight gain monitoring—Regular monitoring of weight gain and growth rates indirectly provide information about the body condition of pigs. Sudden changes in weight may indicate alterations in nutritional status.

4.5.2.9 Index

In order to create an index, Charrette et al. (1996) suggested combining evaluations (tail setting, spinous process of the thoracic vertebrae) with linear measures (pelvic height, width across the rear ham). However, this index wasn't further included in any publications.

These objective methods offer advantages such as precision and reduced variation among individuals. However, the Renco lean meter and other equipment's are relatively expensive, and their use necessitates skilled personnel.

Selecting the appropriate method depends on factors such as available resources, the scale of the operation, and the desired level of precision. Visual scoring methods are often practical for routine on-farm use, while more advanced techniques like ultrasonography and CT scans may be employed in research or specialized settings.

4.6 Salient Features of Sow Condition Scoring (Jackson and Cockcroft 2007)

Maintaining optimal body condition in sows (Table 4.10) is crucial, and this is typically lost during lactation but should be regained in preparation for the subsequent pregnancy. Ideally, sows should enter the farrowing house with a body condition score (CS) of 3.5. A CS lower than 3.0 is considered too low, while a CS higher than 4.0 is deemed too high. At the point of weaning, an acceptable CS is 2.5, and any scores below this indicate a potential issue. Throughout pregnancy, it is recommended that the sow gains approximately 1 CS, ensuring she never falls below CS 2.0 at any stage.

Monitoring trends in sow CS within a unit can be accomplished by assessing different groups of sows at various stages of the reproductive cycle. Boars, on the other hand, should maintain a CS of 3. Sows should undergo condition scoring at weaning, service, mid-pregnancy, and farrowing. Weaning often results in the sow having the lowest CS due to the demands of lactation. A poor CS can lead to an extended weaning to estrus interval and may progress to thin sow syndrome, with young sows in their first or second parity being particularly vulnerable. Delaying mating to the second heat in sows with very poor condition has been shown to increase litter size. Although service typically occurs shortly after weaning, some young sows may experience a significant loss in CS during this period due to preferential muscle growth. Individuals or groups identified as being at risk should receive additional supplementation to achieve target CSs for the next farrowing. Comparing CSs at mid-pregnancy to weaning and service CSs provides insight into the success of remedial feeding strategies (Table 4.11). Suboptimal CS at this stage

Table 4.10 Body condition score guidelines for critical points in the reproductive cycle

Critical point	Target body condition score
Sows at farrowing	3–3.5, with 80% scoring a 3.0
Lactating & weaned sows	2.5–3.5
Remedial action needed	Below 2.5
Middle of gestation	3.0
At breeding	2.5–3.0

Source: NFACC Code of Practice for the Care and Handling of Pigs 2014

4 Body Condition Scoring of Pigs

Table 4.11 Feed adjustments required based on backfat thickness and BCS score

BCS 1	BCS 2	BCS 3	BCS 4	BCS 5
<13 mm Backfat, backbone, rib cage, hip bone prominent sharp to touch	15 mm Backfat, hollowness at loin (flat or "slab" sides). Cavity around tail setting	18–20 mm Backfat, tube shaped, "lean but fit" look	23 mm Backfat, thickening of trunk behind front legs and in neck region. Rear rounded	+25 mm Backfat, excessively thickened trunk behind front legs and in neck region. Bulbous shape
Feed adjustments required				
+ 0.5 kg/day	+ 0.2 kg/day	–	−0.2 kg/day	−0.5 kg/day

can still be corrected before farrowing, serving as a guide to the effectiveness of the chosen feeding regimen. While poor CSs recorded at farrowing cannot be rectified, identifying the issue allows for modifications and improvements in future feeding practices.

4.7 Important Practical Considerations for BCS Assessment and Its Utilization in Sows

1. Enhance the accuracy of condition scores by having two individuals score sows, and then average the resulting scores. Consistency is crucial, so it's advisable for the same individuals to consistently assess the sows in a team approach.
2. Regularly compare your own assessments with those of your colleagues or seek the opinion of an experienced third party, such as a farm veterinarian.
3. Utilize a hands-on approach for scoring, using the palm of the hand, and resort to visual assessment when tactile scoring is not feasible. Emphasize that relying solely on visual inspection may not accurately reflect the sow's condition.
4. Assess sows by examining various locations, including shoulders, ribs, backbone, and hips.
5. Visual and physical condition scoring stands out as the preferred method for evaluating sow condition. Additionally, consider alternative methods like using body condition score measuring tapes.
6. For optimal results, condition score sows at mating and at least twice more between breeding and farrowing. Integrating condition scoring with routine activities like pregnancy checks and vaccinations can save time.
7. A typical schedule involves scoring sows at mating, on day 30 post-mating during pregnancy checks, and around 80 days after breeding.
8. Score sows on a scale of 1–5, allowing for the use of half scores within the mid-ranges. Pigs with a bodily condition score of 1 or lower should not be loaded or transported.
9. Score sows at key points during the reproductive cycle, including weaning and service, midway through gestation and pre-farrowing, and throughout lactation.
10. Ensure feed levels are appropriate and make adjustments as needed. Diets should be formulated to meet protein and energy requirements, accounting for

the specific needs of gilts and young sows to support body lean gain and maintain good overall body condition.

4.8 Summary and Conclusions

Throughout this chapter, we have delved into the different methods employed for BCS, ranging from visual and tactile assessments to more objective tools such as calipers, ultrasound, and image analysis. The integration of these methodologies offers a spectrum of options for swine producers, researchers, and veterinarians to choose the most suitable approach based on their specific goals, resources, and level of precision required.

The significance of accurate body condition assessment cannot be overstated, particularly in the context of optimizing reproductive performance, managing herd health, and implementing effective feeding strategies. The ability to quantitatively measure backfat thickness and evaluate muscle condition contributes to a nuanced understanding of a pig's overall well-being.

The comprehensive exploration of body condition scoring (BCS) in pigs within this book chapter underscores its paramount importance in swine management and production. The evaluation of a pig's body condition serves as a vital tool for informed decision-making across various aspects of husbandry, nutrition, and reproductive management.

In the broader context of animal welfare and ethical husbandry practices, the implementation of robust BCS protocols emerges as a cornerstone. By facilitating early detection of deviations from optimal body condition, producers can proactively address nutritional deficiencies and health concerns, thereby fostering an environment conducive to pig well-being. Keeping sows in the correct condition contributes to enhanced sow longevity and promotes a more consistent reproductive performance.

Continued collaboration among academia, industry professionals, and policymakers is imperative to refine existing methodologies, integrate emerging technologies, and ensure the seamless adoption of BCS practices in diverse swine production systems. Ultimately, the knowledge amassed within this chapter contributes to the holistic advancement of pig farming practices, where the well-being and productivity of the animals are paramount.

In summary, body condition scoring is a valuable management tool that allows pig farmers to make informed decisions about nutrition, reproduction, and overall herd health. It contributes to the efficient and responsible care of pigs, positively impacting both the welfare of the animals and the economic sustainability of pig farming operations.

As we conclude this chapter, it is evident that ongoing research and advancements in technology will continue to refine and enhance our ability to assess and manage body condition in pigs. This knowledge, coupled with practical applications and a commitment to continuous improvement, is crucial for the sustainable and efficient production of pork. By integrating the insights presented in this chapter into everyday practices, swine producers can strive towards achieving optimal herd health, reproductive efficiency, and overall profitability in the challenging and dynamic landscape of modern swine farming.

References

Carrel CL, Huntington PJ (1988) Body condition scoring and weight estimation of horses. Equine Vet J 20:41–45

Charrette R, Bigras-Poukain M, Martineau GP (1996) Body condition evaluation in sows. Livest Prod Sci 46:107–115

Coffey RD, Parker GP, Laurent KM (1999a) Assessing sow body condition. Ky Coop Ext Serv Rep No. ASC-158. http://www2.ca.uky.edu/agcomm/pubs/asc/asc158/asc158.pdf. Accessed 17 Jan 2024

Coffey RD, Parker GR, Laurent KM (1999b) Assessing sow body condition. University of Kentucky, College of Agriculture, Cooperative Extension Service, Lexington

Edmonson AJ, Lean IJ, Weaver LD, Farver T, Webster G (1989) A body condition scoring chart for Holstein dairy cows. J Dairy Sci 72(1):68–78

Evans DG (1978) The interpretation and analysis of subjective body condition scores. Anim Prod 26:119–125

Hutu I, Onan G (2020) Body Condition Scoring. Agroprint, Timişoara, p 10.13140/RG.2.2.18922.67524

Iwasawa T, Young MG, Keegan TP, Tokach MD, Goodband RD, DeRouchey JM, Dritz SS (2004) Comparison of heart girth or flank-to-flank measurements for predicting sow weight. Kansas Agric Exp Station Res Rep 10:17–22

Jackson PG, Cockcroft PD (2007) Handbook of pig medicine. Elsevier, Amsterdam

Knauer MT, Baitinger DJ (2015) The sow body condition caliper. Appl Eng Agric 31(2):175–178

Knudson BJ, Moser RL, Cornelius SG, Pettigrew JE (1985) Estimation of body fat in sows (abstract). J Anim Sci 61(Suppl. 1):104

National Farm Animal Care Council (2014) Code of practice for the care and handling of pigs

Neary M, Yager A (2002) Body condition scoring in farm animals. Purdue Extension service. Purdue University AS-550-W, West Lafayette

Patience JF, Thacker PA (1989) Swine nutrition guide. Prairie Swine Centre, Saskatoon, p 260

Renggaman A, Choi HL, Sudiarto SI, Alasaarela L, Nam OS (2015) Development of pig welfare assessment protocol integrating animal-, environment-, and management-based measures. J Anim Sci Technol 57(1):1–11

Swantek PM, Crenshaw JD, Marchello MJ, Lukaski HC (1992) Bioelectrical impedance: a nondestructive method to determine fat-free mass of live market swine and pork carcasses. J Anim Sci 70:169–177

Whay HR (2008) On-farm monitoring of pig welfare edited by a Velarde and R Gccrs (2007). Published by Wageningen academic publishers, PO Box 200, NL-6700 AE Wageningen, The Netherlands. 203 pp paperback (ISBN 978-90-8686-025-8).€ 49, US $65. Anim Welf 17(1):93–93

Young M, Aherne F (2005) Monitoring and maintaining sow condition. Adv Pork prod 16:299–313

Pig Production and Livelihood Security

Saroj K. Rajak, Satish Kumar, Jaya Bharati, Anil Kumar, Kumar Shambhu Sharnam, and Divya Rani

Abstract

Piggery is one of the important animal husbandry activities and plays critical role in sustaining livelihood security of pig farmers in India. The sale of pigs and value-added pork products contributes significantly to the financial stability of smallholder farmers, empowering them economically and reducing dependence on single-crop agriculture. The capital investment required to establish a pig farm is small and ensures profit in a production cycle of less than a year. With the change in food habits and increasing urbanization, the demand for pork and value-added pork products is expected to increase, which projects greater opportunities in pig rearing. Piggery sector has thus started attracting youths and entrepreneurs as business venture on large scale. This sector is also confronted by challenges like emerging disease outbreaks, limited number of quality germplasm, scarcity of feed and waste disposal issues which need to be addressed on priority basis. Capacity building of farmers on scientific pig husbandry practices and credit availability for capital investment can help in creating supportive environment for piggery sector. Since, pig farming is mostly done by the socio-economically weaker sections of the society, it has larger impact in bringing

---

S. K. Rajak (✉) · D. Rani
Department of Veterinary & A. H. Extension Education, Bihar Veterinary College, Patna, Bihar, India

S. Kumar · J. Bharati
ICAR-National Research Centre on Pig, Guwahati, Assam, India

A. Kumar
Department of Veterinary Clinical Complex, Bihar Veterinary College, Patna, Bihar, India

K. S. Sharnam
Animal & Fisheries Resources Department, Government of Bihar, Patna, Bihar, India

© The Author(s), under exclusive license to Springer Nature Singapore Pte Ltd. 2024
T. Rana, B. Soto-Blanco (eds.), *Good Practices and Principles in Pig Farming*, Livestock Diseases and Management,
https://doi.org/10.1007/978-981-97-4665-1_5

economic prosperity. Pig farming thus has immense potential to ensure the nutritional security as well as enhance the economic status and help in attaining the sustainable development goals.

Keywords

Piggery · Pork · Production · Management · Nutrition · Diseases

5.1 Introduction

Pig husbandry plays a significant role in nutritional and livelihood security of farmers in India, particularly the socio-economically weaker sections and tribal population, who are traditionally engaged in pig farming (Das et al. 2021; Mohakud et al. 2020). Pork is a valuable source of essential nutrients and protein, especially in areas with limited access to other kinds of animal protein. Pig farming, therefore, plays a significant role in boosting nutritional security and treating malnutrition challenges. India has a wide variety of pig farming systems, from traditional free-range scavenging system to technologically intensive commercial pig rearing system. However, the majority of piggery sector is characterized by smallholder, low-input, demand-driven production systems, which are often unorganized in nature (Mohakud et al. 2020; Kadirvel et al. 2023). They often form a part of mixed agriculture or integrated farming system, wherein pigs are reared in backyard system and fed with crop and livestock residues (Kumaresan et al. 2009; Haldar et al. 2017). The existence of different system of pig husbandry practice demonstrates the sectors adaptability to a range of agricultural conditions existing in diverse agro-ecological zones of India (Kadirvel et al. 2023). There are 9.06 million pig population in India, out of which 8.17 million pigs are reared in rural India, which forms about 90.12% of total pig population (20th Livestock Census 2019, DAHD). This signifies the importance of piggery sector in building socio-economic backbone of rural India. Pigs comprise only 1.69% of total livestock population of India as shown in Fig. 5.1, which constitutes approx. 2% of Asia's pig population and 1% of global pig population (FAO 2020). Pork production is limited in India, which represents only 9% of the total animal meat sources in India (NAP 2017). There exists a demand-supply gap in India, and hence pork is imported, mostly in the form of processed meat. This indicates that piggery sector has not been exploited to its full potential in India, which may be mainly attributed to social taboo associated with pigs being reared in free-range scavenging system, giving non-hygienic appeal to pig husbandry. However, with the advent of globalization and modernization of agriculture and allied sector, the sector is gaining steady impetus towards commercialization and upholds great scope for enterprise development in coming years.

Pig farming is widespread across India, with notable concentrations in the north-eastern and eastern region of country; the highest number of pigs exists in Assam (23.17%), followed by Jharkhand (14.12%), Meghalaya (7.79%), West Bengal (5.96%) and Chhattisgarh (5.85%). Pork is the most preferred meat of north-eastern India and there is a high demand for pork and pork products in states of Nagaland,

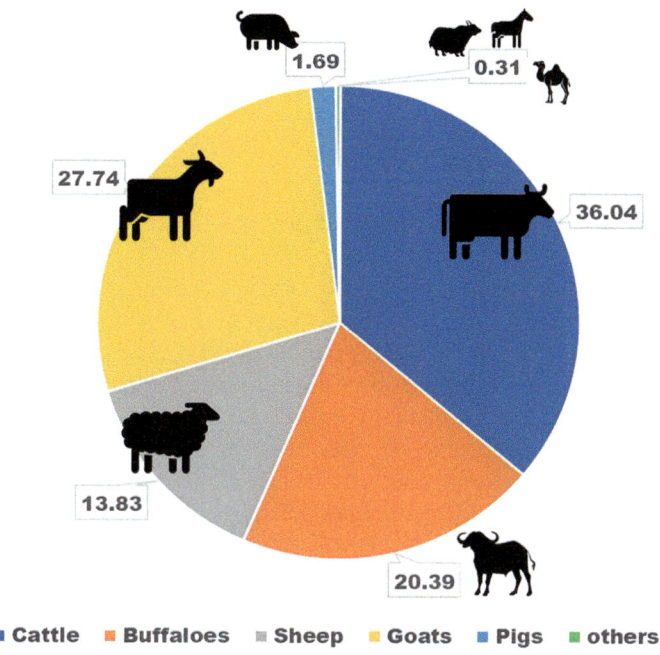

Fig. 5.1 Percentage share of different livestock population in India. (Source: NAP 2017)

Mizoram, Meghalaya, Arunachal Pradesh, Tripura, Sikkim, Assam and Manipur (NAP 2017), accounting for majority of pork consumption compared to other parts of the country. In fact, piggery forms a key livestock species of the farmers in the north-east region, which constitutes 46.80% of India's total pig population. The NE region contributes 19.77% of total pork production in country while the pork consumption is much higher (68.75%) in this region. It reflects the importance of pork in the diet of NE people (Mahajan et al. 2015). The region is enriched in pig biodiversity, covering home tract of 8 out of 14 registered pig breeds of India. Indigenous pigs are treasured for their delicacy and customary value in their distinct geographical area and forms basic constituent of the integrated livestock-agriculture system in the region, particularly among the tribal communities (Kadirvel et al. 2023). Nevertheless, due to urbanization and changing food habits, there also exists substantial demand in urban areas and metropolitan cities throughout the country. Other states, such as Punjab, Haryana, Bihar, Uttar Pradesh, Karnataka and Maharashtra, have come up with intensified pig production hubs, which contribute significantly to the high demand in the north-eastern states and the adjoining regions. Though majority of Indian pigs are reared in subsistence-based smallholder production system, comprising of 2–7 pigs (Kumaresan et al. 2009), the piggery sector is witnessing a major facelift, with greater interest and investments in large-scale commercial pig production system by agri-preneurs (Bharati et al. 2022). The sector is transforming with the scientific, technological and credit inputs from research institutes, government and private sectors. In fact, commercial pig farming and production of pork and value-added pork products is vetted as successful and lucrative venture in India (Thomas et al. 2021). This chapter explores the various aspects of development of Indian pig farming sector, their role in livelihood security, current state, challenges and possible future.

5.2   Pig Genetic Resource of India

India is endowed with a large genetic diversity in livestock population and pig species is not any exception. India has diverse population of pigs and till now 14 breeds of pigs have been characterized and registered as breed. Apart from these 14 breeds, a substantial population of nondescript pigs are also present which still has to be characterized. Indian pig population constitutes 29.40% of descript pig population comprising of 2.80% of exotic, 18.10% cross-bred and 8.20% indigenous pigs, while 70.60% are nondescript pigs, which are yet to be categorized for their breed status. The distribution of different pig population in India is shown in Fig. 5.2.

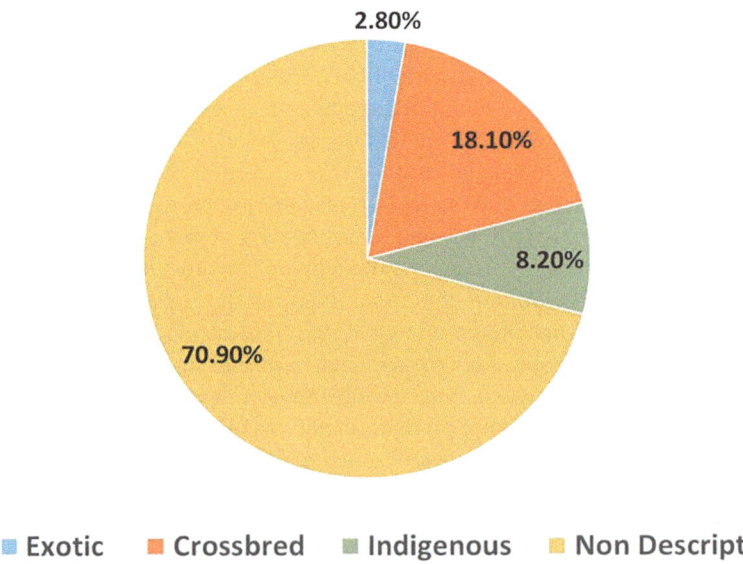

Fig. 5.2 Percentage share of different pig population in India. (Source: NAP 2017)

5.2.1 Indigenous Pig Breeds

Indigenous (breeds or nondescript) pigs are generally smaller in size and their production parameters such as birth weight, litter size at birth and weaning, weaning weight, feed conversion efficiency and average daily gain, are very less. However, these pigs are well adapted to Indian climatic conditions, low input system and possess high disease tolerance. These local pigs are preferred in many parts of the country mainly due to low input requirement, meat quality, taste of pork and their tolerance to the diseases. Some native pigs sexually mature at an exceptionally early age and possess good mothering ability, which directly influence the survivability of their piglets during younger age. The efforts to its conservation and selection to improve the economic traits have been taken up (Das et al. 2024a; Mohan et al. 2023). There are 14 Indian breeds of pigs registered by the ICAR-National Bureau of Animal Genetic Resources. These indigenous pig breeds have evolved by many generations of selective breeding in their breeding tract by farmers and breeders. These indigenous pig breeds are generally lower reproductive and growth performance but well adapted to local climatic conditions and less susceptible to disease. The Ghoongroo pigs found along Indo-Nepal border, Jalpaiguri and Cooch Behar in West Bengal and North Assam is considered as most prolific breed of the country with litter size ranging from 8 to 18. The animals are mostly black coloured with typical bull dog face appearance (Bharati et al. 2023). Other pig breeds include Niang Megha, Agonda Goan, Tenyi Vo, Nicobari, Doom, Zovawk, Ghurrah, Mali, Purnea, Banda, Manipuri black and Wak Chambil. The detailed information on these breeds with their home tract in presented in Table 5.1.

Apart from the 14 well-characterized registered breeds of India, there are several nondescript pig populations having sufficient number but still needs to be characterized as a distinct breed. The Ankamali pigs are found mainly in the state of Kerala along with Karnataka and Tamil Nadu. These pigs are small-sized, compact body, long face with tapering snout and predominantly black coloured with rusty grey and white patches. The slaughter weight is only 37 kg at 8 months (Naskar et al. 2013). Dome pigs are found in Tripura having black to grey coloured with thick bristles from neck to trunk. The adult body weight is about 50 kg (Das et al. 2018). Jovaka pigs is found in Mizoram and Manipur. Animals have small, compact body with long legs and weigh around 40–50 kg at maturity. Golla pigs are reared by the Golla community of Odisha. These pigs are medium-sized with average adult weight of 50–80 kg (Naskar et al. 2013). Lepchamoun pigs are found in Sikkim. The characteristic feature of these pigs is medium-sized with drooping type and adult weight of 80–120 kg (Naskar et al. 2013). Similarly Burudi, Pondi/Jhinga pigs also remain to be characterized.

5.2.2 Exotic Pig Breeds

Due to the poor production performances of indigenous breeds and nondescript pigs, several exotic pig breeds like Large white Yorkshire, Landrace, Hampshire,

Table 5.1 Registered Indigenous pig breeds of India with accession number and their home tracts

Sl. No.	Name of pig	Home track	Accession number
1	Ghoongroo	West Bengal	INDIA_PIG_2100_GHOONGROO_09001
2	Niang Megha	Meghalaya	INDIA_PIG_1300_NIANGMEGHA_09002
3	Agonda Goan	Goa	INDIA_PIG_3500_AGONDAGOAN_09003
4	Tenyi Vo	Nagaland	INDIA_PIG_1400_TENYIVO_09004
5	Nicobari	Andaman & Nicobar	INDIA_PIG_3300_NICOBARI_09005
6	Doom	Assam	INDIA_PIG_0200_DOOM_09006
7	Zovawk	Mizoram	INDIA_PIG_2700_ZOVAWK_09007
8	Ghurrah	Uttar Pradesh	INDIA_PIG_2000_GHURRAH_09008
9	Mali	Tripura	INDIA_PIG_1900_MALI_09009
10	Purnea	Bihar and Jharkhand	INDIA_PIG_0325_PURNEA_09010
11	Banda	Jharkhand	INDIA_PIG_2500_BANDA_09011
12	Manipuri black	Manipur	INDIA_PIG_1200_MANIPURIBLACK_09012
13	Wak Chambil	Meghalaya	INDIA_PIG_1300_WAKCHAMBIL_09013
14	Andmani	Andaman & Nicobar	INDIA_PIG_3300_ANDAMANI_09014

Source: National Bureau of Animal Genetic Resources, India. https://nbagr.icar.gov.in/en/registered-pig/. Accessed 4 Feb 2024

Duroc, Large black are also found which were imported from different countries to enhance the production potential of Indian pigs. These pigs were also used for genetic upgradation of nondescript pigs, by using them in cross-breeding programmes. To incorporate the good characters of both indigenous and exotic pigs, several cross-bred varieties have been developed in the country. Large white Yorkshire (LWY) is the most commonly used exotic breed in India. They have white coat colour with occasional black pigmented spots, erect ears, snout of medium length and dished face. The mature body weight ranges from 300 to 500 kg. Landrace breed is typically white coloured with black skin spots. They have a long body, large drooping ears and long snout. Mature body weight ranges from 250 to 350 kg. Berkshire pigs have white patches on feet, snout and tail. Small head, face depressed in middle and saucer shaped body with flexible ribs are typical characteristics of this breed. Mature body weight ranges from 280 to 350 kg. Hampshire animals are black with white strip across forelegs to shoulder. Typical characteristics include small and erect ears, small and compact body. Sows have good mothering ability. This breed is also used widely in India. The details of exotic pig breed used for cross-breds recommended for different regions of India is presented in Table 5.2.

Table 5.2 Cross-bred pig varieties developed at ICAR-National Research Centre on Pig and different centres of All India Coordinated Research Project (AICRP) on Pigs

Cross-bred varieties	Developed at state	Breeds used for cross-breeding		Exotic inheritance (%)
		Boar	Sow	
Rani	Assam	Hampshire	Ghoongroo	50
Asha	Assam	Duroc	Rani	75
Mannuthy white	Kerala	LWY	Local pigs of Kerala	75
TANUVAS KPM gold	Tamil Nadu	LWY	Local pigs of Tamil Nadu	75
Lumsniang	Meghalaya	Hampshire	Niang Megha	75
HDK-75	Assam	Hampshire	Doom	75
Landly	UP	Landrace	Ghurrah	75
SVVUT 17	Telangana and Andhra Pradesh	LWY	Local pigs of Andhra Pradesh	75
Jharsuk	Jharkhand	Tamworth	Local pigs of Jharkhand	75

Source: Banik et al. (2023) and NRCP (2019)

5.2.3 Cross-Bred Pig Varieties

To enhance the production potential of indigenous breeds of pigs, cross-breeding programmes were initiated in India during 1970–1971 under AICRP (All India Coordinated Research Project) on pig. There are altogether nine cross-bred varieties of pigs developed in different agro-climatic regions of country with better litter size and growth parameters. These cross-bred pigs have inherent quality of exotic breeds like better growth and reproductive performances, superior meat quality and indigenous pigs like high disease resistance and adaptability to local climatic conditions. Moreover, these cross-breds can be reared in medium input system with maximum profit. The details of cross-bred pig varieties developed in different parts of India under AICRP is presented in Table 5.3.

5.3 Current Scenario of Piggery Sector in India

In India, pig farming has long been an essential aspect of rural life, with a long history of activities ingrained in gastronomic and cultural customs, especially in the north-eastern states (Sangtam et al. 2022). In India, backyard systems and small size farms with less than ten pigs are frequently used to raise pigs, which constitutes 65% of total pigs reared in the country (Thomas et al. 2021). In this system indigenous pigs or locally adapted cross-breds are mainly kept in small farms and reared on kitchen waste, green roughages, locally available vegetables, fruits, different kinds of crop stalks and little grains. Though low input is required, the feed conversion efficiency and pork yield are low and takes more time to attain maturity and market weight, sometimes more than 300 days. They are either kept for household consumption or sold in local rural market. Such system provides families a reliable

Table 5.3 Cross-breds to be propagated in different region of country

Sl. No.	Region	Cross-bred recommended
1	Northern India	Large White Yorkshire
		Large White Yorkshire cross
		Landrace cross
2	North-eastern India	Hampshire cross
		Large White Yorkshire cross for Mizoram and Tripura
		Tripple cross with Duroc as terminal sire
		Large Black cross
3.	Eastern India	Hampshire cross
		Tamworth cross (Jharkhand)
4.	Central India	Landrace cross
		Large White Yorkshire cross
5	Southern India	Large White Yorkshire cross
		Tripple cross with Duroc as terminal sire
6	Western India	Large White Yorkshire cross

Source: National Guidelines for formulation of State Pig Breeding Policy. 2019. https://www.dof.gov.in/sites/default/files/2019-12/National%20Guidelines%20for%20formulation%20of%20State%20Pig%20Breeding%20Policy%20.pdf. Accessed 28 Jan 2024

supply of meat and a source of liquid money, especially against the agricultural failures (Bharati et al. 2022). Thus, pig farming has immense potential to ensure nutritional and economic security for the weaker sections of the society. Due to existing gap in demand and supply and to satisfy the rising demand for pork-based products, the backyard farming approach has changed over time and farmers have started embracing contemporary technologies and scientific pig farming managerial techniques. Greater number of pig farmers are interested in getting trained on pig husbandry practices and are shifting to scientific pig rearing (NRCP 2022). The backyard pig farming is transforming to medium-sized local community farms (specialized household), comprising of 50–200 pigs, which constitutes 30% of the total pig production system in India and this has substantially increased to greater than 250% during last decade (Thomas et al. 2018). This change has been brought by progressive farmers, educated youths as well as women-centred rural households, who were aware of gains from piggery and most of them got trained on scientific pig farming. Exotic or cross-bred pigs are raised scientifically, mostly on compounded feed along with locally available feed resources, processed hotel/kitchen waste, resulting in good feed conversion efficiency, which reduces maturation interval and attains optimum slaughter weight (NRCP 2020). The housing, health, vaccine and welfare of pigs are taken due care, which ultimately yields good profit to the farmers. In the coming years, the number of medium-size pig farms are expected to increase manyfold, due to the felt need and economic returns from pig farming. Also, the growth of intensive large sized modern commercial farms with herd size of greater than 200 pigs have been witnessed in the recent past in peri-urban areas, nevertheless, the number of industrialized pig farms is limited to 5% of total pig production in India (Thomas et al. 2020). This system uses highly prolific exotic or cross-bred pigs fed on grain based compounded feed with scientific management practices, skilled labour and proper waste disposal system. Like other domestic stock animals, pigs have a significant economic value that people are aware of these days, which has led to establishment of large intensive pig farms in lesser pork consuming states, like Haryana, Punjab and Karnataka. These states mainly cater to the high demand of pork in the north-eastern states of India. Commercial pig farming can be considered as one of the most successful and lucrative industries in India. This sector will also employ a sizable number of people in a variety of jobs, from off-farm occupations like marketing, processing and transportation to on-farm duties like breeding, feeding and healthcare. Thus, pig farming has great potentiality to emerge as a crucial source of income for rural households as well as agri-entrepreneurs.

5.4 Strength of Piggery Sector

Piggery offers quick economic gains because of its certain inherent traits like early maturity, high litter size, short generation interval, good feed conversion efficiency, and greater dressing percentage (Bharati et al. 2022). A pig is sexually mature in 6 to 8 months (breed dependent) and can be bred at 8 to 9 months of age (Banik et al. 2019). Sows (female pig) can give birth to 6 to 16 piglets, depending on the breed reared, in one farrowing. Due to short gestation period (114 days), a sow can farrow twice in a year, which can result in 12–32 piglets per year, which can be sold at good market price (Banik et al. 2020). The piglets grow quickly and reaches market weight of average 80 kg at 8 months of age (Banik et al. 2021). Pigs have high feed conversion rate and thus it can gain more body weight than other livestock with per kg of feed consumed (Reyes-Palomo et al. 2023). Pigs have greater meat yield due to higher dressing percentage and thus fetch decent income on sale of meat (Banik et al. 2022). Pig fat has varied industrial use in manufacture of soap, poultry feed and many chemicals. Pig manure can be used as fertilizer for crops, kitchen gardens and fish pond (Sulabh et al. 2017). Pigs form a perfect component of integrated farming system in rural India (Kumaresan et al. 2009).

Indian piggery sector is dominated by smallholder farming system, employing household labour, wherein women form an important component. They are managed on low-inputs or constitutes a part of integrated farming system at subsistence level agriculture. As a part of pig-fish, pig-fish-poultry, pig-fish-crop or pig-fish-poultry-crop integrated farming practice, piggery forms an integral part of livelihood security component for rural communities.

The pig fish integrated farming is very successful model of integrated farming. In this system, the pig shed is constructed on the pond embankment or over the pond in such a way that the wastes are directly drained into the pond that serves as feed for fish. The pig manure acts as excellent pond fertilizer and raises the biological productivity of the pond and consequently increases fish production. Fishes feed directly on the pig excreta which contains 70 percent digestible food for the fish and no additional feed is necessary for fish which normally accounts for 60 percent of the total input cost in conventional fish culture. The water of pond can be used for bathing pigs and cleaning the pig shed. This system can also be integrated with other livestock component like goat and cattle along with the crops, which can increase the net utility and income from the integrated farming system. The schematic diagram of an integrated farming system with pig as a major component is shown in Fig. 5.3. Integrated farming has special significance as it can improve the socio-economic status of weaker rural communities, especially the tribals who traditionally raise pigs and can take up fish-pig farming easily (Tripathi and Sharma 2001). Piggery thus provides livelihood diversification to farmers due to its ease of integration with other animal husbandry and agricultural system along with minimal inputs for pig farming practice. These smallholder integrated production systems have contributed immensely to livelihood and nutritional security, poverty reduction and women empowerment in rural areas (Fang 2022).

There exist two types of pig breeder farm functioning in India based on the ownership. The different state government operated pig breeder farms consisting of 10–50 sow units operate on a single-site farrow-to-finish farm system and supply pigs to the local farms. These breeder farms supply high quality germplasms of exotic breed like Large White Yorkshire, Landrace, Hampshire, Duroc along with the Indigenous breeds like Ghoongroo (Thomas and Sarma 2017). In addition to these State Governments pig farms, 15 farm units exists under AICRP and six farm units under Mega Seed Project (MSP) in different states of India harbouring varied agro-climatic conditions, which is monitored by nodal institute ICAR- National Research Centre on Pig, located at Guwahati, Assam, the state with highest pig population. The cross-bred varieties developed in these farm units are propagated to the farmers field with adequate technical and scientific support. The other type is privately owned pig farms, which fulfil majority of pork demand in the country.

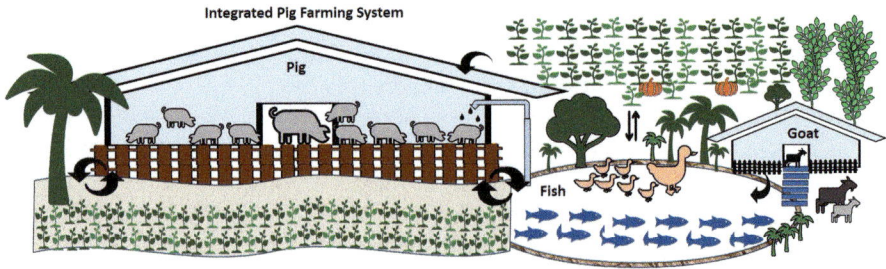

Fig. 5.3 Schematic representation of integrated pig farming

5.5 Major Challenges and Their Mitigation Strategies in the Piggery Sector

The pig population in India had a positive growth rate until the year 2000, except during 1960s. However, since last three Livestock Census of India (2007, 2012 and 2019) a decrease in the pig population in India was found (Indian Livestock Census 2019). This may be due to sever reasons like incidences of diseases, greater slaughter percentage and reduced repopulation rate, which may be attributed to many challenges faced by the piggery sector in India. Although a significant percentage of rural and tribal population depend on pig rearing for their livelihood, the majority don't have means to undertake scientific pig farming with improved foundation stock, proper housing, feeding, disease control and farm management practices, mainly due to low capital resources. These gap needs to be filled to develop piggery and pork-based industries. The development of pig industry is hampered by disease outbreak which leads to mass culling of pigs, as happened with outbreak of PRRS and African Swine Fever. Several steps have been taken to revive and support the sector, so as to harness the maximum benefits from it. Some of them may be described as follows:

5.5.1 Incidences of Diseases

The advancement of pig farming in India has been significantly impeded by the emergence of various pig diseases and infectious ailments. Over the past decade, the pig industry has experienced the onslaught of two devastating diseases: PRRS (Porcine Reproductive and Respiratory Syndrome) and ASF (African Swine Fever). These diseases have led to substantial pig mortality and have adversely affected pig farming. ASF, in particular, is a highly contagious viral swine disease that inflicts significant economic losses on pig farmers. All types of pigs, including exotic, cross-bred, and indigenous breeds, are susceptible to this virus, except Doom pigs (Das et al. 2024b). The direct losses attributed to ASFV in India are estimated to be INR 2.76 billion (US$ 37.32 million) (Mohan et al. 2021). The extensive culling measures employed by farmers to control these diseases have resulted in considerable losses, causing many to abandon pig farming.

To combat these diseases, maintaining stringent biosecurity measures on the farm and implementing effective farm management practices to reduce disease incidence are crucial. This includes thorough cleansing and disinfection, accurate disease diagnosis and reporting, and timely vaccination against major economic diseases. In addition to PRRS and ASF, other diseases such as Classical Swine Fever, Foot and Mouth Disease, porcine parvovirus disease, coccidiosis, respiratory diseases, *Streptococcus suis* infection, and parasitic diseases also contribute to the challenges faced by pig farming (Kumar et al. 2023). Parasitic infestation, in particular, can depress growth rates and feed efficiency by up to 10%. The most noteworthy ectoparasite causes serious economic bearing on growing pig is sarcoptic mange. Besides ectoparasites, intestinal parasites like *Ascaris suum*, which is a pig nematode significantly hinders the growth of young pigs and decreases the profitability of pig farms. Proper vaccination schedule, deworming, health-check-up, monitoring for disease outbreaks and biosecurity measures can help prevent occurrence of deadly diseases in a pig farm and this avoid production losses.

5.5.2 Quality Germplasm

There are many pig breeds available in India, however, pig farming is still concentrated on the local nondescript pigs that are easily available in the area. The major difficulty in pig development is the acute shortage of breeding boars and sows. There is shortage of quality germplasm due to limited number of breeder stock. With the aim to address the issue of availability of good quality germplasm to the farmers field, the MSP on Pig was launched in 2008 by the Indian Council of Agricultural Research. The MSP have the objective to produce high-quality agroclimatically adapted swine germplasm at scientifically managed farm units and distribute this superior germplasm to farmers. Under MSP, each centre must have to produce 900 piglets in a year. Currently there are six MSP centres at Ranchi, Guwahati, Sikkim, Nagaland, Kerala and Tripura producing and distributing piglets of Jharshuk, HDK-75, HDK-75, Rani, Mannuthy White and LWY breeds/varieties, respectively.

5.5.3 Shortage of Good Quality Feed

Feed costs accounts for 65–75% of total investments in pig production chain. Quality pig feed is a major problem, causing low productivity index in pigs with respect to potential growth rates and mature body weight. The cost of commercially available feed grains for livestock is quite high and India is facing a shortage of feed supply, which is mainly dependent on good rains, temperature and change in cropping preferences. Climate change, water scarcity, drought and floods hamper the cost and supply of livestock feeds, which in turn adversely affects the cost of production. High pig feed cost is one of the major deterrent to good economic returns on pig farming, since in the absence of quality feed, even the best germplasm will not yield optimum production performance, which in turn decreases the profitability of a pig farm. Cost-effective alternative feeding technologies incorporating locally available nonconventional feed resources into pig feed can decrease the input cost of pig production. To achieve this target, ICAR- National Research Centre on Pig has developed technology for economic feed formulation using locally available feed ingredients, which has greatly reduced the feeding cost without affecting the nutrient utilization.

5.5.4 Waste Disposal

Pig farming results in the generation of a substantial amount of organic waste, including faeces, urine, wastewater, food waste, and slaughter by-products. If the solid and liquid waste produced by pig farms is not managed with precision, it poses potential hazards to humans, animals, and the environment (Solís-Tejeda et al. 2021). The environmental consequences encompass the release of greenhouse gases and the eutrophication of water bodies, contributing to adverse effects (Hollas et al. 2022). Moreover, there is a heightened risk of zoonotic diseases affecting humans and other domestic animals (Holt et al. 2016). To mitigate these risks, it is imperative to ensure proper waste disposal practices in pig farming. Diseases such as cysticercosis, influenza, tuberculosis, and leptospirosis can be transmitted through the contamination of food items with pig waste. While pig waste can serve as valuable manure to enhance soil fertility and provide nutrients to plants, without adequate treatment, it becomes a potent source of zoonotic pathogens. The methods of waste disposal includes returning to soils, Aerobic biological treatment Anaerobic biological treatment and drying process of dung. The sustainable growth of the pig industry necessitates the implementation of regulations governing the proper treatment and disposal of pig waste to safeguard human health, animal well-being, and the overall environment (Ström et al. 2018).

5.5.5 Social Taboo

In India, pig farming had a negative reputation in society and only socially backward and tribals used to raise pigs. But due to increase in urbanization, education and awareness on demand of protein source of food as well as change in the of food preferences and habit of Indian people, pig farming is no longer restricted to the poor and backward community. With the increasing institutional and credit support, now, many entrepreneurs and youths see pig farming as a business opportunity due to its ability to produce cheap source of protein in relatively short duration. In the north eastern region of the country, where it was mainly reared by the tribal people, the non-tribals have also come forward with the interest and investment in large scale commercial pig farming and slaughter house facilities. The change in social behaviour of people have increased the acceptability of pig farming and pork products in the country.

5.6 Opportunities in Pig Husbandry

Pork constitutes only 4.98% of total meat produced in India and there exists demand and supply gap in pork and pork products in our country. In the North-East (NE) states, pig is the most important livestock and pork is the meat of choice, hence, there exists felt demand for pork and pork products. India is a net importer of pork, which indicates the growth potential of this sector in the domestic market.

Interestingly, majority of the country's pork is consumed in NE India which is sourced from own production as well as procurement of pigs from other parts of the country. With the increase in urbanization and change in food habits, the social taboo associated with consumption of pork is wanning and its demand is felt in cities in hotel, restaurant and institutional sectors which is projected to increase in near future. Moreover, the opening of NE corridor of India to South East Asian countries would provide a global market for pork and pork products. Looking into the domestic and export potential, piggery sector has great promises to uplift rural economy, however interventions are required at every level of pig production to exploit this sector at its full potential. Hence, pig husbandry in India requires an immediate transformation from backyard subsidiary enterprise to a commercial venture. Aaddressing the gap in the pork supply through domestic production, would fulfil the twin objective of economic and nutritional security. For the purpose, interventions through innovations in science and technology customized to piggery sector is the need of hour which can bring revolutionary change in pig husbandry and brand it a lucrative enterprise.

5.7 Technology-Led Interventions for Upscaling Piggery Sector

In India, major share of piggery exists in unorganized sector, which deserves science and technology-driven support to make it a full-fledged enterprise. Piggery has a great potential to develop as an enterprise, but it also has many challenges, which should be considered and prior measures should be taken to tackle bottlenecks of this sector. Major constraints include the emerging and re-emerging pig diseases, availability of high-quality germplasm, availability of pig feed, limited access to market and lack of skilled labour. Smallholder pig farmers are less likely to adopt safe pork production practices due lack of access to commercial slaughter house. Coming years will bring additional challenges with respect to shrinking land availability, global warming, water scarcity and emerging diseases. In such scenario, use of technology in pig husbandry can maximize economic gains, minimize the environmental impact and make piggery more sustainable.

5.7.1 Integrated Health Management

An integrated health management is required in the piggery sector, which is often ravaged by infectious pig diseases like PRRS, CSF, ASF, porcine endemic diarrhoea (PED), etc. which poses a serious threat to the pig farming. Since, these diseases cause high mortality and morbidity and incurs huge expenditure on treatment, culling and repopulation of a farm, strict biosecurity measures should be maintained and proper health calendar should be followed in a pig farm. In fact, safeguarding the piggery sector requires effective disease management strategies, including vaccination campaigns, awareness among farmers and prompt veterinary services. To control the occurrence of disease in a pig farm, technology for disease diagnosis, vaccine production, disease forecasting, real time health monitoring, warning system for breach of biosecurity and effective management of the farm is required. The institute has developed rapid diagnostics kits for surveillance and monitoring of major pig diseases. Application of advanced scientific technology in managing farm records, breeding, feeding, vaccination and culling schedule will facilitate early identification of diseases and precise individual treatments to improve herd performance, reduce the use of antibiotics and contribute to public safety.

5.7.2 Artificial Insemination

In order to fulfil germplasm demand, technologies towards infrastructure development for semen banks and artificial insemination (AI) in pigs can help in bringing superior genetics and reduce the cost of keeping breeding boars at farm (Kadirvel et al. 2017). AI technology developed at ICAR-National Research Centre on Pig, has helped in dissemination of quality germplasm to rural interiors where boars of superior merit are not available. Another area to increase the pig productivity is through use of superior germplasm adapted to particular agro-climatic region. The cross-bred pigs have better growth and reproductive performances, superior meat quality, high disease resistance and adaptability to local conditions. Moreover, these cross-breds can be reared in medium input system with maximum profit.

5.7.3 Availability of Abattoir and Access to Market

The availability of modern abattoir assures clean and hygienic slaughter practice, which in turn reduces odour and noise arising out of slaughter to the neighbourhood residents. Technologies pertaining to mechanized on-rail slaughter system, pork processing, cold chain maintenance and retail outlets is critical to hygienic pork production. In this direction, technology for cost-effective rural pig slaughter house as per FASSI requirements and technology for hygienic transport of pork for local distribution has been developed by ICAR- National Research Centre on Pig. Enhancing the cold storage infrastructure, processing facilities and expanding market access are also essential for hygienic selling of pork and pork products, which warrants good return to the pig farmers. Also, improved market connections for sale of piglets, finishers and culled pigs can guarantee higher prices for pig producers, thus protecting their means of subsistence. Pig value chain linkage and market support are mandatory for sustainable growth of piggery sector.

5.7.4 Housing Management

Innovative technologies for housing management to combat heat stress is critical to successful outcome of pig production system so that piggery sector is least affected by extremes of climate change and continue to be a driver of growth.

5.7.5 Pork Processing and Value Addition

The value addition of the pork can provide pig farmers with maximum economic gains. Towards this end, technologies on pork processing and manufacture of value-added products viz. restructured pork products, enrobed pork products, emulsion-based pork products and self-stable pork products have been developed by our institute. These products have high consumer demand and acceptability.

5.7.6 Capacity Building

Capacity building initiatives, such as training programmes on scientific pig farming practices, disease management and entrepreneurship development are essential for uplifting the piggery sector from smallholder production system to agri-business enterprise. This can be achieved by empowering farmers with knowledge and skills in scientific pig husbandry practices.

5.8 Policy Implications and Credit Support

It is essential to have policies that address disease control, strengthen market connections and provide funding for the construction of infrastructure in the piggery sector. The Government of India have provided a conducive atmosphere with supportive policies for piggery husbandry that can establish a framework that facilitates the sustained expansion of the piggery sector and empowers the socio-economically backward rural poor in uplifting their livelihood.

5.9 Way Forward

India offers a diverse array of pig farming systems, ranging from conventional free-range system to cutting-edge, technologically advanced intensive farming system. These systems' coexistence is a testament to the industry's flexibility in a variety of agricultural environments. In the coming years, the traditional medium holder unorganized pig farms are projected to be gradually replaced by commercial-type, scientifically organized farms. However, majority population may still continue with smallholder pig farming system. Bringing Indian piggery sector at global level will require innovative technology interventions with regards to pig breeding,

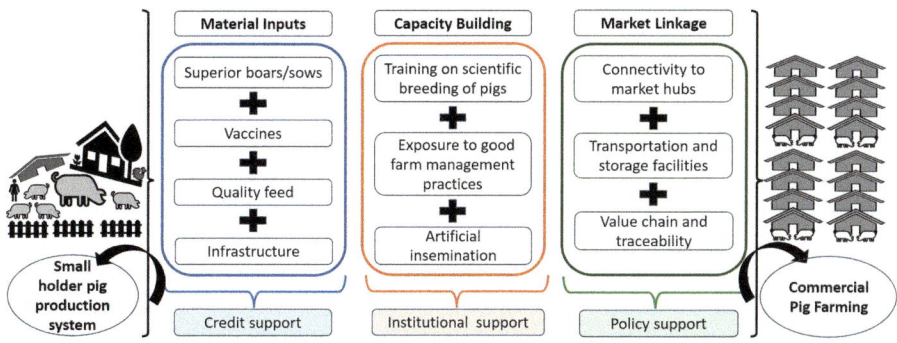

Fig. 5.4 Schematic representation of different level of support system required to upscale pig husbandry in India

feeding, housing, vaccines, pork processing, entrepreneurship development, and financial support. Adoption of scientific pig farming practices by smallholder farmers will ultimately augment profitability, efficiency and sustainability of the overall pig production system. However, credit linkage support and empowerment of the economically poor rural tribal households, especially women are critical to adoption of new technologies for pig rearing. A multi- and trans-disciplinary research collaboration for developing technologies catering to the needs of pig husbandry is an essential component of current and future road-map to development of this sector. Inclusion of awareness programmes and extension activities on scientific breeding, AI, vaccination, safe pork production practices, hygienic slaughter house, value addition and waste management will be critical to successful piggery enterprise in rural India (Vision 2030 2011). Nevertheless, advanced scientific knowledge in pig farming should be integrated with information and communication technologies, which should be simple, cheap and customized for use by farmers so as to reach the remote corner of country and can be easily adopted by semi-literate women and tribal folks engaged in piggery sector. A schematic representation of different level of support system required to upscale pig husbandry in India is shown in Fig. 5.4.

5.10 Conclusion

Pig farming is more than just a customary agricultural practice in India; for many households, it is a lifeline that makes a substantial contribution to economic development and nutritional security. Through tackling obstacles and capitalizing on opportunities, the decision-makers and interested farmers and agri-preneurs can work together to fortify the pig farming industry, guaranteeing the financial stability of many farmers associated with the pig husbandry. The piggery development programmes implemented by Department of Animal Husbandry and Dairying, veterinary and animal husbandry departments of different states and research organizations in India, have led to significant advancement in piggery and many start-ups are coming forward to invest in this sector. ICAR-National Research Centre on Pig, through its technology intervention, various institute-village-linkage programmes and capacity building activities have played a critical role in abridging the knowledge gap and enhancing scientific pig farming practices of farmers and entrepreneurs. Thus, it cannot be doubted that science and technology-led interventions can address the future challenges for growth and development of piggery sector. Technology-oriented pig husbandry will change the adoption behaviour and motivate more farmers and agri-preneurs towards choosing pig husbandry over other livestock. At the same time, a consolidated effort of animal husbandry departments, veterinarians, entrepreneurs, credit institutions, co-operatives, and the entire pork-value chain is essential to integrate innovative technologies in pig husbandry for convincing results. The limitations of the sector should be resolved by adopting good management practices and implementing contingency plan in pig husbandry practices. If the current barriers to their economic upbringing are eliminated, pig farming could significantly contribute to the improvement of this section's socio-economic position. The nation hasn't yet embraced pig farming as a profitable industry and the pork-based food processing industry can witness improved productivity and sustainability, changing the pig farming landscape and creating a diverse and distributed industry.

References

Banik S, Mohan NH, Barman K, Das PJ, Kumar S, Kumar S, Sahu AR (2019) Annual report all India coordinated research project and mega seed project on pig (2018–19). ICAR NRC on Pig, Guwahati

Banik S, Mohan NH, Kumar S, Barman K, Das PJ, Kumar S, Yadav AK, Ahmad SF (2020) Annual report all India coordinated research project and mega seed project on pig (2019–20). ICAR NRC on Pig, Guwahati

Banik S, Mohan NH, Kumar S, Das PJ, Barman K, Kumar S (2021) Annual report all India coordinated research project and mega seed project on pig (2020–21). ICAR NRC on Pig, Guwahati

Banik S, Mohan NH, Kumar S, Das PJ, Barman K, Kumar S (2022) Annual report all India coordinated research project and mega seed project on pig (2021–22). ICAR NRC on Pig, Guwahati

Banik S, Mohan NH, Kumar S, Das PJ, Barman K, Kumar S (2023) Annual report all India coordinated research project and mega seed project on pig (2022–23). ICAR NRC on Pig, Guwahati

Bharati J, De K, Paul S, Kumar S, Yadav AK, Doley J, Mohan NH, Das BC (2022) Mobilizing pig resources for capacity development and livelihood security. In: Agriculture, livestock production and aquaculture: advances for smallholder farming systems, vol 2. Springer International Publishing, Cham, pp 219–242

Bharati J, Kumar S, Mohan NH, Das BC, Devi SJ, Gupta VK (2023) Ovarian follicle transcriptome dynamics reveals enrichment of immune system process during transition from small to large follicles in cyclic Indian Ghoongroo pigs. J Reprod Immunol 160:104364

Das S, Naha BC, Saini BL (2018) Adopted way of pig rearing practices in Tripura. J Entomol Zool Stud 6(4):1673–1678

Das G, Hajra D, Mukherjee R, Hembram S, Roy B (2021) Sustainable income generation of the farmers through pig farming: a case study in Terai region of West Bengal. J Livestock Sci 12(4):241–245

Das PJ, Kumar S, Choudhury M, Banik S, Pegu SR, Kumar S, Deb R, Gupta VK (2024a) Characterization of the complete mitochondrial genome and identification of signature sequence of Indian wild pig. Gene 897:148070

Das PJ, Sonowal J, Sengar GS, Pegu SR, Deb R, Kumar S, Banik S, Rajkhowa S, Gupta VK (2024b) Characterization of African swine fever virus outbreak in India and comparative analysis of immune genes in infected and survived crossbreed vs. indigenous Doom pigs. Arch Virol 169(7):145

Fang J (2022) Intensified smallholder pig farming in rural Yunnan: implications for livelihood, culture, gender, health and environment. Open J Soc Sci 10(8):55–67

FAO (2020) Food and Health Organization. FAOSTAT 2020

Haldar A, Das D, Santra A, Pal P, Dey S, Das A, Rajkhowa D, Hazarika S, Datta M (2017) Traditional feeding system for pigs in Northeast India. Int J Livestock Res 7(8):122–132

Hollas CE, do Amaral KGC, Lange MV, Higarashi MM, Steinmetz RLR, Barros EC, Mariani LF, Nakano V, Kunz A, Sanches-Pereira A, de Martino Jannuzzi G (2022) Life cycle assessment of waste management from the Brazilian pig chain residues in two perspectives: electricity and biomethane production. J Clean Prod 354:131654

Holt HR, Inthavong P, Khamlome B, Blaszak K, Keokamphe C, Somoulay V, Phongmany A, Durr PA, Graham K, Allen J, Donnelly B (2016) Endemicity of zoonotic diseases in pigs and humans in lowland and upland Lao PDR: identification of socio-cultural risk factors. PLoS Neglect Trop Dis 10(4):e0003913

Kadirvel G, Bujarbaruah KM, Kumar S, Ngachan SV (2017) Oestrus synchronization with fixed-time artificial insemination in smallholder pig production systems in north-East India: success rate and benefits. S Afr J Anim Sci 47(2):140–145

Kadirvel G, Devi YS, Naskar S, Bujarbaruah KM, Khargariah G, Banik S, Singh NS, Gonmei C (2023) Performance of crossbred pigs with indigenous and Hampshire inheritance under a smallholder production system in the Eastern Himalayan hill region. Front Genet 14:1042554

Kumar S, Bhushan B, Kumar A, Panigrahi M, Bharati J, Kumari S, Kaiho K, Banik S, Karthikeyan A, Chaudhary R, Gaur GK (2023) Elucidation of novel SNPs affecting immune response to classical swine fever vaccination in pigs using immunogenomics approach. Vet Res Commun 48:941–953

Kumaresan A, Bujarbaruah KM, Pathak KA, Das A, Bardoloi RK (2009) Integrated resource-driven pig production systems in a mountainous area of Northeast India: production practices and pig performance. Trop Anim Health Prod 41:1187–1196

Livestock Census (2019) Livestock census, India. Department of Animal Husbandry, Dairying and Fisheries, Ministry of Agriculture, Govt. of India

Mahajan S, Papang JS, Datta KK (2015) Meat consumption in North-East India: pattern, opportunities and implications. J Anim Res 5(1):37–45

Mohakud S, Hazarika R, Sonowal S, Bora D, Talukdar A, Tamuly S, Lindahl J (2020) The extent and structure of pig rearing system in urban and peri-urban areas of Guwahati. Infect Ecol Epidemiol 10(1):1711576

Mohan NH, Misha MM, Gupta VK (2021) Consequences of African swine fever in India: beyond economic implications. Transbound Emerg Dis 68:3009–3011

Mohan NH, Pathak P, Buragohain L, Deka J, Bharati J, Das AK, Thomas R, Singh R, Sarma DK, Gupta VK, Das BC (2023) Comparative muscle transcriptome of Mali and Hampshire breeds of pigs: a preliminary study. Anim Biotechnol 34:3946–3961

NAP (2017) National action plan on pig. Department of Animal Husbandry and Dairying, Govt. Of India. https://dahd.nic.in/sites/default/filess/NAP%20on%20Pig%20.pdf. Accessed 3 Feb 2024

Naskar S, Niranjan SK, Banik S (2013) Utilization of pig genetic resources in India. In: Pundir RK, Niranjan SK, Behl R (eds) Sustainable utilization of indigenous animal genetic resources. NBAGR, Karnal, pp 120–125

NRCP (2019) Annual report of ICAR-NRC on pig. ICAR-National Research Centre on Pig, Rani, Guwahati

NRCP (2020) Annual report of ICAR-NRC on pig. ICAR-National Research Centre on Pig, Rani, Guwahati

NRCP (2022) Annual report of ICAR-NRC on PIG. ICAR-National Research Centre on Pig, Rani, Guwahati

Reyes-Palomo C, Aguilera E, Llorente M, Díaz-Gaona C, Moreno G, Rodríguez-Estévez V (2023) Free-range acorn feeding results in negative carbon footprint of Iberian pig production in the dehesa agro-forestry system. J Clean Prod 418:138170

Sangtam HM, Laskar SK, Thomas R, Das A (2022) Physico-chemical and sensory attributes of traditional pork products incorporated with Anishi at refrigerated storage (4±1° C) under vacuum packaging. J Anim Res 12(5):667–673

Solís-Tejeda MÁ, Lango-Reynoso F, Castañeda-Chávez MDR, Ruelas-Monjardin LC (2021) Analysis of the environmental impact generated by backyard swine production in Tepetlán. Pig Production Impact in Tepetlán, Veracruz Agro Productividad, Veracruz

Ström G, Albihn A, Jinnerot T, Boqvist S, Andersson-Djurfeldt A, Sokerya S, Osbjer K, San S, Davun H, Magnusson U (2018) Manure management and public health: sanitary and socioeconomic aspects among urban livestock-keepers in Cambodia. Sci Total Environ 621:193–200

Sulabh S, Shivhare P, Kumari A, Kumar M, Nimmanapalli R (2017) Status of pig rearing in India. Status of pig rearing in India. Int J Vet Sci Anim Husbandry 2(3):30–32

Thomas R, Sarma DK (2017) Pig production and pork processing—Indian perspective. Jaya Publishing House, New Delhi

Thomas R, Rajkhowa S, Pegu SR (2018) Package of Practices (PoP) on organized backyard pig production for pig bondhus. ICAR-National Research Centre on Pig, Rani, Guwahati

Thomas R, Mohan NH, Rajkhowa S (2020) Creation of entrepreneurship through public private partnerships. Indian Farming 70(1):40–42

Thomas R, Singh V, Gupta VK (2021) Current status and development prospects of India's pig industry. Indian J Anim Sci 91(4):255–268

Tripathi SD, Sharma BK (2001) Integrated fish-pig farming in India. In: Integrated agriculture-aquaculture: a primer (No. 407). Food and Agriculture Organization, Rome, pp 54–55

Vision 2030 (2011) National Research Centre on Pig. ICAR-National Research Centre on Pig, Rani, Guwahati

Pig Behavior and Welfare

Subir Singh

Abstract

This chapter highlights the significance of understanding pig behavior for effective management and welfare practices. Pigs exhibit a diverse range of behaviors influenced by factors like natural instincts, environmental conditions, social interactions, and overall health. From foraging and nursing behaviors to responses to stressors, studying pig behavior provides valuable insights into their wellbeing, enabling farmers to create environments that promote positive behaviors and minimize stress. Moreover, the sensory capacities of pigs, including vision, smell, taste, and hearing, play a crucial role in their perception of the environment and social interactions. Normal behavioral patterns, such as farrowing and nursing behavior, resting, eating, drinking, and excretory behavior, are essential for understanding pigs' biological functions and aiding in their care in commercial production settings. The chapter also discusses abnormal behaviors observed in production conditions, such as piglet savaging, belly-nosing, oral, nasal, and facial behaviors, and tail biting, which serve as indicators of poor welfare and health. Additionally, behaviors exhibited by sick and compromised pigs are outlined, emphasizing the importance of prompt individual care to prevent adverse impacts on welfare. Furthermore, the chapter addresses the concept of swine welfare, emphasizing the need to ensure good health, safety, comfort, nutrition, and freedom to express natural behaviors for pigs. It discusses the World Organization for Animal Health's recognition of Brambell's Five Freedoms as pivotal guidelines for swine welfare, covering physical, mental, and natural aspects. Finally, the chapter delves into the assessment of swine welfare through

S. Singh (✉)
Department of Veterinary Medicine and Public Health, Faculty of Animal Science, Veterinary Science and Fisheries, Agriculture and Forestry University, Rampur, Chitwan, Nepal
e-mail: ssingh@afu.edu.np

© The Author(s), under exclusive license to Springer Nature Singapore Pte Ltd. 2024
T. Rana, B. Soto-Blanco (eds.), *Good Practices and Principles in Pig Farming*, Livestock Diseases and Management,
https://doi.org/10.1007/978-981-97-4665-1_6

animal-based criteria, including behavior, morbidity and mortality rates, changes in body weight and condition, reproductive efficiency, physical appearance, handling responses, lameness, and complications from common procedures. Overall, the comprehensive understanding of pig behavior and welfare indicators presented in this chapter aids in promoting the well-being of pigs in diverse production settings.

> **Keywords**
>
> Pig behavior · Welfare practices · Environmental conditions · Sensory capacities · Abnormal behaviors · Swine welfare

6.1 Introduction

Understanding pig behavior is essential for effective pig management and welfare. Pigs exhibit a wide range of behaviors that are influenced by their natural instincts, environmental conditions, social interactions, and overall health. From foraging and nursing behaviors to social interactions and responses to stressors, studying pig behavior provides valuable insights into their well-being and allows farmers to create environments that promote positive behaviors and minimize stress. By delving into the intricacies of pig behavior, farmers can improve husbandry practices, enhance productivity, and ensure the overall welfare of these intelligent and complex animals (ASEAN 2015).

6.2 Sensory Capacity of Pigs

Pigs possess remarkable sensory abilities, including vision, smell, taste, and hearing.

- **Vision**: Pig eyes have a diameter of about 24 mm, akin to human eyes, with a total optical power estimated at 78 diopters. They have approximately 310° vision and can detect colors, particularly blue and green-yellow.
- **Smell**: Pigs have a keen sense of smell, vital for identifying piglets, recognizing pen-mates, and dominant individuals. This sense is utilized in swine production, for instance, in suppressing aggression among pigs.
- **Taste**: Similar to humans, pigs can discern between sweet, sour, salty, and bitter tastes. They favor sweet, meaty, and cheesy flavors and reject bitter food. Sweeteners can encourage newly weaned piglets to consume solid feed.
- **Hearing**: Pigs have a hearing range similar to humans but are more sensitive in the ultrasonic range. They use audio cues for communication, such as nursing grunts from sows to synchronize nursing in a room.

6.3 Normal Behavioral Patterns

Normal behavior in pigs, observed under natural conditions, supports biological functions such as survival and reproduction. While commercial rearing environments differ from natural conditions, understanding these behaviors aids in improving animal care in indoor commercial production.

- **Farrowing and Nursing Behavior**: Sows typically spend 2–4 h giving birth, with nursing being continuous initially and later becoming rhythmic. Piglets can successfully suckle milk within 45 min after birth.
- **Resting Behavior**: Pigs spend a significant portion of their time resting or lying down, adjusting their posture based on thermal comfort.
- **Eating and Drinking Behavior**: Pigs eat multiple meals per day, with eating events peaking in the morning and evening. Drinking behavior usually occurs within 10 min of eating, with the frequency influenced by environmental factors such as temperature.
- **Excretory Behavior**: Pigs tend to excrete in cool, wet areas and lie in warm, dry areas. They prefer to excrete away from feeding areas and may create dunging areas near drinkers.

6.4 Abnormal Behaviors and Possible Causes Under Production Conditions

Abnormal behaviors, not observed under natural conditions, serve as indicators of poor welfare in pigs under production settings. These behaviors can compromise production performance or health.

- **Piglet Savaging**: Occurs when gilts or sows kill their piglets after birth, often associated with factors such as lighting conditions.
- **Belly-Nosing**: A repetitive, non-functional behavior typically observed in early weaned piglets, influenced by stress associated with early weaning.
- **Oral, Nasal, and Facial Behaviors (ONF)**: Stereotypic behaviors often seen in pregnant sows, associated with factors like limit feeding and barren environments.
- **Tail Biting**: Damaging abnormal behavior more often observed in grow-finish pigs, linked to various factors including malnutrition and discomfort.

6.5 Behaviors in Sick and Compromised Pigs

Sick pigs typically display behaviors such as reduced feeding, drinking, and interactions with pen mates, along with increased resting, huddling, and shivering. Prompt individual care is essential for sick and compromised pigs to prevent adverse impacts on their welfare.

Understanding the sensory capabilities and normal behavioral patterns of pigs facilitates better management practices, promoting their welfare and overall well-being in swine production.

6.6   Swine Welfare

Animal welfare pertains to how well an animal adapts to its farm environment. A high level of welfare is achieved when the animal enjoys good health, resides in a safe and comfortable space, receives proper nutrition, and can freely engage in natural behaviors. Swine welfare, specifically, ensures that pigs do not endure pain, fear, or distress. This entails effective disease management, balanced nutrition, humane handling, suitable shelter, and ethical slaughtering practices (WOAH 2019a, b).

Since 1979, the World Organization for Animal Health has recognized Brambell's Five Freedoms as pivotal guidelines for addressing swine welfare needs, encompassing physical, mental, and natural aspects:

1. Freedom from physical and thermal discomfort.
2. Freedom from injury, disease, and pain.
3. Freedom from distress and fear.
4. Freedom to express normal behavioral patterns.
5. Freedom from malnutrition, thirst, and hunger.

6.7   The Swine Welfare Predicament

Swine welfare has emerged as a pressing issue, prompting ethical and practical considerations, particularly at the farm level. It's crucial to distinguish between the biological and ethical dimensions of animal welfare. Historically, farmers focused on meeting the biological needs of animals to maximize production, enhancing facilities, environments, feed quality, and herd health. However, in recent years, consumers have begun questioning the ethical implications of animal production methods, seeking transparency in food production. With increased exposure through social media, consumers recognize animals' capacity to feel pain and suffering and perceive a disparity between pigs' behavioral needs and their living conditions. Consequently, consumers advocate for changes in practices like castration, tail biting, and ear tagging. Addressing these concerns promptly is essential for farmers to remain competitive in a global market where maximizing production is no longer the sole priority.

6.8 Welfare of Pigs

Animal-based criteria for assessing the welfare of pigs involve measurable outcomes directly observable in the animals themselves. These indicators serve as tools for evaluating pig care management in alignment with welfare principles. The selection and interpretation of these indicators depend on various factors such as regional practices, herd health, pig breed, and environmental conditions. Key indicators for pig welfare include behavior, morbidity rates, mortality and culling rates, changes in body weight and condition, reproductive efficiency, physical appearance, handling responses, lameness, and complications from common procedures.

Behavior: Behaviors observed in pigs serve as crucial indicators of their welfare status.

Stereotypy: Stereotypies, repetitive behaviors resulting from frustration or central nervous system dysfunction, can signal poor welfare. Examples include sham chewing, stone chewing, tongue rolling, teeth grinding, bar biting, and floor licking. Although stereotypies generally suggest welfare issues, their presence may not always correlate directly with stress.

Apathy: Apathetic behavior, characterized by reduced activity and lack of interest or emotion, indicates compromised welfare.

Agonistic Behavior: Agonistic behaviors, exhibited in conflict situations, encompass offense, defense, and submissive or escape actions. These behaviors may involve contact (e.g., biting, pushing) or non-contact threats (e.g., body postures, gestures).

Play Behavior: Play behavior, marked by neuroendocrinological responses and a sense of enjoyment, aids in preparing pigs for unexpected situations by enhancing their adaptability and coping skills.

Certain behaviors, such as play behavior and specific vocalizations, may indicate good welfare and health status in pigs. Conversely, altered locomotory behavior, agonistic behavior, apathetic behavior, and stereotypies may signify welfare or health issues.

6.9 Morbidity Rate

Morbidity rates, encompassing infectious and metabolic diseases, lameness, and post-procedural complications, serve as indicators of herd-level welfare. Monitoring morbidity rates can help identify potential welfare problems, with factors like mastitis, metritis, leg problems, respiratory issues, and reproductive diseases being of particular concern.

6.10 Mortality and Culling Rates

Mortality and culling rates impact the overall welfare of the herd, reflecting factors affecting the animals' productive life span. Regular recording and analysis of mortality and culling rates, along with necropsy examinations, aid in understanding and addressing welfare issues.

6.11 Changes in Body Weight and Condition

Deviation from expected growth rates or significant fluctuations in body weight and condition among pigs can indicate compromised welfare and health status. Such changes may warrant further investigation into potential welfare concerns and reproductive efficiency.

6.12 Reproductive Efficiency

Reproductive performance serves as an indicator of animal welfare and health. Poor reproductive outcomes, including low conception rates, high abortion rates, and small litter sizes, may signal underlying welfare problems.

6.13 Physical Appearance

Physical attributes such as body condition, presence of ectoparasites, abnormal texture or hair loss, skin lesions, and discharges can provide insights into the welfare status of pigs. Abnormalities in posture, feet, and legs, as well as signs of emaciation or dehydration, may also indicate compromised welfare.

6.14 Handling Responses

Responses to handling procedures can reveal the level of fear and distress experienced by pigs. Indications of poor welfare may include avoidance of handlers, excessive vocalization, and injuries sustained during handling.

6.15 Lameness

Lameness and gait abnormalities in pigs may stem from various factors, including genetic predisposition, diseases, nutrition, and environmental conditions. Scoring systems are available to assess lameness and gait abnormalities, with affected pigs experiencing difficulties in accessing feed and water and potentially suffering pain and distress.

6.16 Complications from Common Procedures

Proper performance of procedures such as castration, tail docking, and vaccination is crucial to avoid compromising animal welfare and health. Indicators of post-procedure complications may include infection, lameness, behavioral changes indicative of pain or distress, increased morbidity and mortality rates, and reduced feed intake.

In summary, a comprehensive understanding of these animal-based indicators is essential for assessing and promoting the welfare of pigs in various production settings.

References

ASEAN (2015) ASEAN GAHP on animal welfare and environmental sustainability module: layers, broilers and ducks. ASEAN, Bangkok

WOAH (2019a) Terrestrial Animal Health Code, Chapter 7.1 Introduction to the recommendations for animal welfare

WOAH (2019b) Terrestrial Animal Health Code, Chapter 7.13 Animal welfare and pig production systems

Pig Exhibition Rules, and Its Monitoring

7

Jessy Bagh, Annada Das, Kaushik Satyaprakash, and Tanmoy Rana

Abstract

Pig exhibitions or pig shows are the events where pigs are judged for their general health, breeding characteristics, conformation, meat quality using the art and science of visual assessment. These exhibitions give a platform for breeders to interact with others and publicize better genetic makeup among different breeds and farmers. Pig exhibition rules are essential components to ensure ethical and efficient showcasing of pigs during the show. The regulations typically encompass criteria for pig health, welfare, breed standards, and exhibitor conduct. Monitoring involves regular inspections by qualified veterinarians to assess compliance with the set guidelines. These inspections aid in safeguarding the well-being of exhibited pigs, maintaining breed integrity, and upholding exhibition standards. Additionally, stringent record-keeping and transparency measures often accompany monitoring efforts to track health histories, breeding data, and exhibitor practices. Overall, effective rules and monitoring frameworks promote

J. Bagh
Department of Livestock Production and Management, College of Veterinary Sciences and Animal Husbandry, OUAT, Bhubaneswar, Odisha, India

A. Das
Department of Livestock Products Technology, Faculty of Veterinary and Animal Sciences, WBUAFS, Kolkata, West Bengal, India

K. Satyaprakash
Department of Veterinary Public Health and Epidemiology, Faculty of Veterinary and Animal sciences, RGSC, BHU, Mirzapur, Uttar Pradesh, India

T. Rana (✉)
Department of Veterinary Clinical Complex, West Bengal University of Animal and Fishery Sciences, Kolkata, West Bengal, India

© The Author(s), under exclusive license to Springer Nature Singapore Pte Ltd. 2024
T. Rana, B. Soto-Blanco (eds.), *Good Practices and Principles in Pig Farming*, Livestock Diseases and Management, https://doi.org/10.1007/978-981-97-4665-1_7

pig exhibition practices and contribute to the overall advancement of pig farming sector.

Keywords

Pig · Exhibition · Rules · Monitoring · Breed · Health

7.1 Introduction

The pig exhibitions showcase various pig breeds, emphasizing their unique characteristics, utility, and cultural significance. These events can promote breed standards of recognized breeds including Large White Yorkshire, Hampshire, Berkshire, Poland China, Chester White, Hereford, and Tamworth etc. for any agricultural education and industry advancements (Lauterbach et al. 2019; Woods et al. 2019). Exhibitors present pigs based on specific criteria such as health, conformation, and genetic purity. Through competitions, demonstrations, and educational sessions, pig exhibitions foster knowledge exchange among breeders, farmers, and the public. Additionally, these events often highlight sustainable farming practices, economic potential, and conservation efforts related to pig farming.

The other names for the pig exhibition are "pig show" and "swine exhibition" (Bowman et al. 2016; Brophy et al. 2024). Pig shows are the events where pigs are judged for their quality. They are evaluated based on a variety of things including composition (muscle vs. fat), spaciousness and skeletal integrity along with the general appearance and adapt body structure to the respective breed. Owners can buy or raise their pigs as a companion animal. The body composition of a pig can be changed by simply changing its feed. A pig show is often part of a larger agricultural show. During the show in addition to pigs, cattle, goats and sheep can also be shown at the agricultural shows. Pig shows begin at farms where the owner and other pig farmers meet and select the pig with the best breeding characteristics. After that, the breeders meet at a designated location and bring an outside scientific person as judge to evaluate their pigs and rank them. A show is usually divided into classes so that the judge can study less pigs at once. Typically, these classes are based on the breed of pig, weight or age.

The exhibition of pigs has a rich history globally. Historically, pigs have been showcased in various contexts, from agricultural fairs to cultural events. These exhibitions or events have highlighted the breeding advancements, conformation for meat quality, role in sustainable agriculture and rural livelihoods and even as pets. Over the years, pig exhibitions in the global context have evolved to reflect changing societal values, agricultural practices, and conservation efforts. Currently, these pig exhibitions continue to play a vital role in fostering knowledge exchange, promoting breed diversity, and advancing the pig industry.

The 4-H pig project, operated by a youth development organization (4-H) in the United States of America (USA) is administered by the cooperative extension system, a partnership between National Institute of Food and Agriculture (NIFA) and the United States Department of Agriculture (USDA). In this project, youth aged 9–19 years raise pigs and showcase their pigs at the local exhibitions, which aid in gaining valuable experience in animal husbandry practices (Reynnells 2012; Sarjahani and Harmon 2015).

Several national and international animal expos, fairs, exhibitions, and shows are organized across the globe that also exhibit pigs. Some examples of such events are depicted in Table 7.1.

Pig exhibition rules and their subsequent monitoring represent foundational pillars in ensuring the ethical, humane, and standardized showcasing of the pigs

Table 7.1 A glimpse of national and international animal expos, fairs, exhibitions, and shows are organized across the globe

Sl. No.	Name of the expo/fair/ exhibition/ show	Country	Remarks	References
1.	World Pork Expo	United States of America (USA)	An annual event and is the largest gathering of pork producers and industry professionals. It includes exhibitions, educational sessions, and competitions.	Clark et al. (2020, 2021)
2.	Royal Agricultural Winter Fair	Canada	This event often features competitions for livestock, including pigs. It provides a platform for showcasing different breeds and promoting agricultural excellence.	Bathelt and Spigel (2012)
3.	Melbourne Royal Show	Australia	This show includes competitions for various livestock, including pigs. Winners are recognized for their excellence in breeding mostly.	Darian-Smith and Wills (2001), Ressia et al. (2022)
4.	EuroTier	Germany	This is an international trade fair. It covers various aspects of animal husbandry, including pig farming, and attracts participants and exhibitors from around the world.	Bünger and Schiller (2022)
5.	China Animal Husbandry Expo (CAHF)	China	It is one of Asia's largest exhibitions on latest developments in animal husbandry activities. Also includes a section dedicated to pig farming, involving shows and exhibitions.	Sun et al. (2022)
6.	Indian International Poultry, Agri and Dairy Expo	India	This event focuses on various aspects of agriculture and livestock farming, with sections dedicated to pig exhibitions.	DAHD (2024)
7.	Krishi Mela (Agricultural Fair)	India	These are annual agricultural fairs, where farmers showcase their livestock, including pigs. These events often include competitions and exhibitions.	Gangil et al. (2019), Kumar Rajak et al. (2020)

(Woods et al. 2019). As the pig exhibition serves various purposes, including breed promotion, agricultural education, and industry advancement, it becomes imperative to establish stringent guidelines that prioritize animal welfare, breed authenticity, and exhibitor integrity. These rules often encompass a broad spectrum, ranging from specific health and welfare criteria for pig to detailed standards concerning exhibitor conduct and facilities. By setting clear parameters, such regulations aim to maintain the highest standards of care, prevent potential disease outbreaks, and ensure fair competition among participants.

The monitoring mechanisms often accompany the rules and typically involve systematic inspections conducted by qualified veterinarians and other regulatory body members (Andreoletti et al. 2011; Beltran-Alcrudo et al. 2019). These inspections serve as checkpoints to assess exhibitor compliance with the predefined guidelines, offering insights into areas of improvement and corrective actions where necessary. Furthermore, comprehensive record keeping and transparency measures complement monitoring efforts, facilitating traceability, and accountability within the exhibition environment. In essence, the harmonious interplay between well-defined rules and rigorous monitoring mechanisms fosters an environment of responsibility, credibility, and advancement within the pig exhibition landscape.

7.2 Objectives of a Pig Exhibition

The basic objectives of a swine or pig exhibition are given below:

1. It aims to exhibit the best animals at exhibitions or fairs.
2. Allows individual breeders to exchange ideas and experiences at exhibitions.
3. It provides the opportunity to make comparisons between superior types animals both within and between herds.
4. It offers new breeders the opportunity to get in touch with established breeders and serves as a contact point advertising medium.
5. It helps to publicize and increase better genetic makeup among different breeds, farmers' pride stimulates public interest in the desired species.
6. It builds healthy comparison and encouragement with other farmers.
7. It helps in the revival of the livestock industry.
8. It promotes human well-being through higher yields at high production animals.

7.3 Significant Rules and Its Monitoring to Organize a Successful Pig Show

All exhibitors must guarantee that the animals registered are fit. It is the exhibitor's responsibility to be in good health for the last 6 months and ensure compliance with all current laws. The major rules and regulations pertaining to pig exhibitions or shows are given below:

7 Pig Exhibition Rules, and Its Monitoring

1. The show organizer reserves the right to disqualify or refuse participation of any animal for any reason that they deem unsuitable.
2. Exhibitors should comply with all American Public Health Association (APHA) or any standard international, national and/or local authority rules and regulations, including necessary movement documents and identity (ID) cards.
3. Exhibitors are responsible for insuring all of their property and livestock on the exhibition grounds or at another place belonging to him or for which he is responsible. The fair requires exhibitors to address employer liability insurance if required by law. The show also requires exhibitors who cannot influence liability and product liability. Therefore, all exhibitors are obliged to provide the appropriate equipment insurance coverage.
4. Exhibitors should have an electronic movement licence, i.e. a movement licence to and from the exhibition/show.
5. Owners who wish to exhibit pigs at exhibitions must submit their declaration and application licence to the local authority. The show pigs must be kept apart from the rest of the herd for 21 days. The return of a pig from a show does not trigger the 21-day rule. If the owner wishes to exhibit a pig that has returned from a previous exhibition show and who are in segregation must be filled out completely and resubmitted the declaration that his pig herd continues to adhere to the 21-day rule before the local authority issues another licence. When returning from the show with the subsequent notification, a new 21-day separation or quarantine period begins (Shurson et al. 2023).
6. Owners may sell exhibition pigs that are kept separately if they have the opportunity to do so complete the declaration and obtain a licence from your local authority. Such pigs should be moved with a licence and would be banned for some period in the new premises.
7. Commercial pigs must come to the exhibition with metal ear tags.
8. The Exhibition Committee reserves the right to combine any class where the entries are not enough, i.e. in case of four entries or less.
9. Exhibitors can make an unlimited number of registrations in each class by payment of the relevant fees.
10. Morning unloading—All animals must be on the ground by 8 a.m. Failure to comply may result in the exhibitor being fined or disqualified.
11. Pigsties—The Council reserves the right to limit the allocation of pens for future exhibitions to exhibitors who cannot show the number of pigs registered and included in the livestock programme. Please indicate the number of pens on the transfer form is necessary. All bedding and food must be provided by the exhibitor. Drinking water is available on the exhibition grounds. When If possible, some straw will be provided.
12. Exhibitors/breeders are responsible for exhibiting their pigs in the exhibition area ring.
13. Exhibitors/breeders are responsible for the field marking of their pigs.
14. To comply with pig health regulations, please ensure all animals do so is free of lice.

15. Colouring—The use of artificial dyes, bleaches and/or powders on pigs, as well as removing or adding stains by artificial means is forbidden.
16. Objections or protests—Notice of objection to any entry or decision must be made in writing to the pig secretary as soon as possible and definitely not later than 60 min after completion of the assessment (National Pork Board 2016). It is mandatory to deposit the fees to be made; if the investigation does not uphold the protest, the deposit will be refunded if the exhibition is failed for the means of the show. Any objections will be taken into account by the exhibition secretary with two members of the pig committee.
17. Any exhibitor of an animal that, to his knowledge, is sick with a contagious or infectious disease or with which you have come into contact. Exhibitor pays compensation to the show for any suffering of animal. Furthermore, in the exhibition, veterinarian may certify that each animal was sent in his opinion, such an exhibition representing a possible source of danger be quickly removing from the exhibition grounds by the participant or otherwise treating the animal as per veterinarian's instructions to prevent the spread of infection.
18. The prerequisite for entry is the event of a serious accident of animal as per recommendation of the show veterinary officer. If the animal is slaughtered, every effort will be made to contact the animal owner to ask for consent to slaughter. If the owner is not available within an appropriate time, a local police officer will be asked to take care of this written power of attorney as per Animal Welfare Act 2006. If emergency veterinary treatment is required at the exhibition, the costs of any medications and medicines are at the expense of the exhibitor in the treatment of their animals and payment must be made directly to veterinarian at the time of treatment.
19. Any animal receiving veterinary treatment at the time of exhibition must not be entered unless the owner has had the animal examined by their own veterinarian and it is safe for them to enter the exhibition. Animals, however anyone who continues to show signs of illness will not be allowed to appear.
20. Exhibitors must be aware of and prepared for the evaluation of special prizes for their animals when they are called.
21. In all cases the judge's decision is final.
22. Young handler classes, boars, and sows are not permitted.

7.4 General Considerations for Choosing a Pig for the Exhibition

Given below are the general considerations for choosing a pig for the exhibition:

1. A pig ideally should be at an age of 6–7 month at the time of show or trade fair and weight around 240–270 pounds for improved globally recognized breeds or 60–90 kg in Indian conditions (Tummaruk et al. 2007). Pigs younger than 6 months may not be sufficient at the time of the show. Pigs who are too old may need to be held and may lose their muscle shape.

2. Breed is a personal choice, but mixed breeds will win the show with over 50% of the whole time. However, there are advantages to shows that exhibit by breed. The representation of pure-bred animals, for example, the Duroc or Yorkshire division, may not be so more competitive than the cross-division.
3. Make sure the pig is vaccinated. If not, ask your vet about vaccinating pigs. Pigs should also be marked by ear notching. This is the way breeders identify pigs and are required at major exhibitions.
4. Choose a pig that is large, long, broad, muscular, and runs smoothly. At the moment, judges select pigs with production characteristics such as larger bodies, more bones, more width between the front and back legs and a little more fat on them than a few years ago. The judges want a muscular shape, for example a wide area top and a big butt. Remember that pigs that win are complete pigs, for example the pig with the biggest butt or the longest pig probably won't win the show. The pigs which win are pigs that have a combination of all of these characteristics.

7.5 Basic Requirements for a Pig Show

7.5.1 Animal Welfare

Animal welfare requirements for pig shows prioritize the well-being of pigs. These standards typically include providing adequate space for each pig, ensuring proper ventilation, access to clean water, and appropriate nutrition. Stress reduction measures, humane handling, and veterinary care are essential. Strict adherence to animal welfare guidelines aims to create a positive environment for the pigs, promoting their physical and mental well-being during the exhibition (Elzen et al. 2011; Nielsen et al. 2022).

7.5.2 Biosecurity at the Pig Show

Biosecurity begins when the pig is purchased and is an ongoing process for the lifespan of a pig project. Biosecurity involves working with experts who know how to prevent exposing pigs to diseases and keeping them healthy and developing written development biosecurity plan specific to pig farms and exhibitions/shows (Cochran et al. 2023). Before visiting an exhibition, check the biosecurity plan to protect the health of pigs.

7.5.3 Public Health Concerns

Breach in biosecurity measures, direct contact with the pigs carrying subclinical infections and long-term exhibition durations (3–7 days) are the potential risk factors for outbreak and spread of zoonotic diseases like Influenza A Virus (IAV) or swine flu virus (Vincent et al. 2009; Killian et al. 2013), spread of methicillin resistant *Staphylococcus aureus* (MRSA), African Swine Fever (ASF) (Cochran et al. 2023) etc. (Dressler et al. 2012). Many trade fairs and exhibitions require health papers listing specific vaccines administered (e.g. flu) and may require targeted health testing, i.e. for Porcine Respiratory and Reproductive Syndrome (PRRS) or Porcine Epidemic Diarrhoea Virus (PEDV) (Schweer et al. 2016; Zhao et al. 2021). The owners must bring a written record of the pig's treatments and vaccinations to the show. As part of the biosecurity plan, determining an appropriate vaccination schedule and testing by a veterinarian is required for the pigs taking part in the show.

7.5.4 Equipment

A driving tool is needed to lead or herd your pig. A livestock cane, pig whip, riding crop, pig prod or plastic pipe can be used. A spray bottle of water is used to clean and cool the pig before sending into the show ring.

7.5.5 Showmanship

Showmanship is one of the most important parts of a 4-H pig project (Reynnells 2012; Sarjahani and Harmon 2015). Presentation skills assess the ability to display an animal effectively in a competitive show. Advanced planning and practicing at home are the keys to becoming a good showman and achieving success in gaining the trust of pig. Continuous planning, practice, and a well-groomed appearance at the trade fair will be helpful. When presenting the pig, the exhibitor's personal appearance is important. The exhibitor should dress neat and for safety reasons, should wear leather shoes or boots.

7.5.6 Preparing a Pig Before the Show

For show, fair or exhibition only healthy pigs are taken. Pig health is assessed daily before exhibition for the following (National Pork Board 2016):

- Is the pig eating normally?
- Is the pig coughing, throbbing or having problems in breathing?
- Does a pig have a fever? Does it seem depressed?
- Does the pig have loose stools?
- Does the pig limp or have sores on its feet?

Wash the pig thoroughly and make sure it is clean. Particular attention is to be paid to the ears; scrubbing them with a brush cleans the ears sufficiently. Be particularly careful of not getting water into the pig's ears as this will affect its balance. Before returning clean the pig in the stall, check for dirty bedding, remove it and replace it. Before entering the ring, spray the pig with water from your spray bottle. Brush the pig's hair the way it naturally lies. Do not straighten the pig's torso back. This makes the top appear flat and a flat top appears bold. Comb hair with a natural part along the spine to give the pig a meatier appearance. Do not use oils or powders on a pig as these will make the animal hot. Pigs having oils or powders on them or clipping, are not accepted by packers. Daily brushing and proper hair care can prevent this. Pruning before the show is required. If you suspect your pig is sick, notify the show veterinarian or other fair official immediately. Keep the surrounding area of the pig and equipment clean. Hands are frequently washed with warm, soapy water after contact with pigs or pigs' equipment.

7.5.7 Presenting Pig in the Show Ring

The pig should reach the show ring on time for exhibition. When the pig enters the ring, it may become excited and start running. Quietly walk up to your pig and take control by driving it 10–15 feet in front of judge. Do not drive the pig closer to the judge unless instructed to do so. Using driving stick is wise to move the animal. Pat the animal from their front rib forward (shoulder and jaw area) to rotate. If you want the pig to move the right, touch its left cheek. If the pig needs to move to the left, touch it to the right. Always drive forward by tapping below the rear ankles. Never hit the pig. Keep the tacker away from the top, loin, and ham areas. Never place your hand on the ham or loin of the pig. Always keep the swine between you and the judge; this gives the judge a full view of your pig. When you move or change direction try not to propel the animal with your hands or knees. When walking with the pig, move calmly with the pig while remaining on its side to the judge. A slight bend in the waist may give you better waist control of the pig. If the owner is calm, the pig will be calm and respond to commands. Be polite at all times and be aware of danger zones, i.e. a group of other pigs, noises etc.

7.5.8 Biosecurity Measures on Returning to Farm After the Show

The biosecurity measures to consider when returning home are given below.

Implementing a biosecurity plan doesn't end with the end of a show. It is equally important to take biosecurity precautions when you return home the ones you take before and during the show. Clean, disinfect, and dry all equipment brought to the show including trailer, even if the show was ended. Isolate any returning pigs from the home herd. Work with the veterinarian to determine duration of isolation and for all health testing requirements. Monitor returning pigs daily for signs of illness and notify the veterinarian, if necessary, about the health of the pigs. If the pigs become ill, it is very important to allow them to make a full recovery before they go to another show (Cha et al. 2009; Shurson et al. 2023).

7.6 Raising a Pig for Show

A successful pig project begins with the healthy raising of pigs (von Borell et al. 2020; Hunter et al. 2022). To find the best pig, evaluating a pig from its show ring potential is not enough, but also the health status of the farm from which the pig comes is important. Given below are the things to be taken care of during raising of pigs for show purposes:

(a) For raising pigs from own farm for show purposes, the following things must be taken care of:
   Ensure that purchased or delivered semen and/or breeding material is free of any clinical illness or a "minimal medical history." Maintain facilities separate from the home herd to isolate new and returning animals and enable testing for targeted diseases. Review the vaccination record with a qualified veterinarian.
(b) If pigs are bought for show purpose, the following things must be taken care of:
   Review current herd health history. Inquire whether the current group of pigs has experienced any health problems for sale. If so, were they treated and with what product? Check that the pig has received current vaccinations. When were those vaccinations given? Be sure to provide contact information if any additional question arises.

The selection process should place an emphasis on determining health and health status, and potential performances of the pigs.

7.6.1 Housing of Pigs

Some suggestions for pigsties are given below:

1. The pen should be dry. Muddy stalls can lead to lameness, pigs don't do well in mud and are difficult to work with pig.
2. The pen should be covered and provide protection from heat and sun in summer, cold and rain in winter. White pigs or any pig with patches of white on them will get sunburned, especially after being clipped for show.
3. Provide at least 60 square feet (6 feet × 10 feet) per pig on concrete, twice that space if on sand. About ten pigs must be kept per pen. Pigs kept in small stalls require more exercise (Elzen et al. 2011; Jayaraman and Nyachoti 2017).
4. A play pen is a good idea, but pigs can also be exercised outdoors.
5. Clay and black land soils typically do not make good pens. These floors are hard on the pig's feet and legs.
6. If pigs are kept on concrete, layer them with either wood chips or river sand. These pens need to be cleaned regularly and the chips or sand replaced.
7. Clean fresh water must be available at all times. People often underestimate the amount of water a pig needs.

7.6.2 Feeding a Pig

1. Pigs can self-feed until they weigh 120 pounds, sometimes more depending on how much weight the pig needs to gain and how fat the pig is. It's hard to describe when a pig gets fat, but there are some signs that the pig is getting fat looks round across its torso and ribs, or looks loose or soft on the stomach under its neck (Pomar et al. 2021).
2. Feed should contain at least 16% protein, younger pigs (<120 pounds) require one higher protein content than 16% (Święch 2017).
3. However, nutritional supplements (mineral supplements, vitamin supplements) can be used are probably not necessary with a good feed.
4. After the pig gains 120 pounds, it should be fed about 2 pounds feed twice a day, depending on how much growth is needed. Pigs at this moment should be weighed twice a month. At around 160 pounds, the amount of feed may need to be increased up to 2.5 pounds twice daily (Schneider et al. 2006).
5. Pigs are best to weight once a week when they reach the weight of 200 pounds. Weighing pig before feeding tells more accurate weight.
6. With every weight group, the rate of weight gain must be monitored. When a pig is gaining weight too quickly, a general advie is needed to maintain a certain weight of the pig or slow the increase rate. To check the gain rate there are different way to execute this, i.e. determined by the frame size of the pig and leanness of the pig before weighing the increase gain rate. If a pig is not gaining enough weight, the amount of feed should be adjusted and increased.

7.6.3 Working with the Pig Before the Show

1. To tame the pig is the first step in working with a pig. A pig should come towards you when you enter its pen. When he runs, go to the middle of the enclosure and stand still. As soon as the pig realizes that you are not chasing him, he will become curious and examine what you are. Let it come towards you slowly and then start rubbing it. For a nervous pig, it may take a week or more to get all the rub done.
2. The next step is to teach the pig to follow the commands. A pig show stick or a piece of polyvinylchloride (PVC) pipe is required. Start working with the pig in the stall. Start slowly tapping the pig on the side of its head to encourage it to turn over. Tap is—don't hit. Pigs are intelligent and have good memories. If you hurt the pig, make him do it what you want will be more difficult. To encourage him to walk, tap its ribs. Don't tap it on the back or butt. The pig will take time to learn so have patient.
3. Once the pig has learned a few commands, it is best to take the pig for a walk in a bigger pen. Try to train the pig to stay away from the fence. A good rule is to turn the pig over before it reaches the fence. Pigs often walk besides fences and peep for the openings. If this occurs at the show, it will be very difficult to show your pig and for the judge to see. Get someone to stand in the enclosure and

pretend them as the judge. Try to keep the pig 10–15 feet away from them (National Pork Board 2016). Execute keeping the pig between you and the judge. If your pig runs, approach it calmly. When you run, your pig simply runs more. When your pig gets hot, put it back in its pen and sprinkle water on its nose and feet.

4. Pigs require exercise. A pig that is not exercised, can become fat and gets tired quickly. At a pig show, it will take typically 15–30 min in each class (National Pork Board 2016; Woods et al. 2019). If a pig is not walked at home, it will be tired at exhibition and also very hard to manage. When it is tired, its muscles are also tense, causing the pig to walk stiffly. The animal should be started for the exercise programme slowly, walked for just 5–10 min and then increased the time daily.
5. The pig is washed once a week with adjoin conditioner after washing. It will help to make the pig's hair shine. Washing of the pig sould be started from 2 weeks before the show every 2 to 3 days.

7.6.4 Working with the Pig at the Show

1. It is prime to know how much the pig weighs each day and how much feed and water it drinks and eats. Water makes the biggest difference for the pig's weight. A quart of water contributes to weight about 2 pounds and a pig can drink a gallon (4 quarts) at once. That indicates it would gain 8 pounds. The best plan is to water the pig several times a day with small amounts of water (Manu and Baidoo 2020). A pig that loses weight, it is generally meant that it doesn't get enough water, but a pig that also drinks lot of water at once can become too full and make you look pot-bellied. Even pigs who don't do that will lose some of muscle shape. This is what stands out the most when their top becomes round instead of flat.
2. Pigs need to be trained at the exhibition, the stables are usually small and the pigs become stiff. Usually, walking with them for 5 min two to three times a day is enough to help keep them loose.
3. If it is hot at the trade fair, the animals should be provided a fan. If it's cold, pig shoud be protected from the cold.
4. Make an effort to use the same food, shampoo, and other supplies that you have already used at home. Every time you change the product you may experience a rash or stomach upset.
5. The exhibitors and handlers should stay calm. Pig gets nervous if the persons around are nervous.

7.7 Judging a Pig During the Exhibition

Swine or pig judging is an art and science of visual appraisal or making subjective evaluation of pigs (Wemelsfelder et al. 2009; Holloway and Morris 2014). Swine producers use it for selecting herd sires and gilts for breeding programmes and also for market purpose. Factors required for judging swine or pig are:

(a) **Evaluating degree of muscling**: Best indicator is by looking muscles thickness gain in the centre of hams (Felicioni et al. 2020). Second look is seeing the width between the widest part of the ham while pig is walking or standing. The back legs should be presented since there is a lot of space between them, this wide base helps the pig move things through the ring comfortably. Third look is sharp top line. Muscular top line should have a butterfly shape showing big muscles the pork chops (Fernandes et al. 2020; Murray 2020).

(b) **Growth:** Growth is simply how the judges compare the pigs in the class. Assume that every pig in the class was born at approximately the same time, i.e. the largest and the best muscular pig has the best growth in the class. To ensure that the pig grows well, make sure that the pig builds muscle every day, not just fat. When evaluating a market eater, growth is really important to remember that every pig has the same amount of pork chops, ham, and loins. It is better to use a large pig than a small pig because the large pig simply has more meat (Bosi and Russo 2004). Good body length, correct slope and set of shoulder of a pig are shown in Fig. 7.1.

(c) **Capacity:** Capacity refers to the pig's ability to eat, grow, and reproduce (Kyriazakis 2006). Assess the width (Fig. 7.2), depth (Fig. 7.3), and length of the pig. Look at the width between the legs on the hooves. Does the width between the hooves correspond to the width between legs at chest and thighs? The goal is a consistent long throw width. The further apart the legs are, the

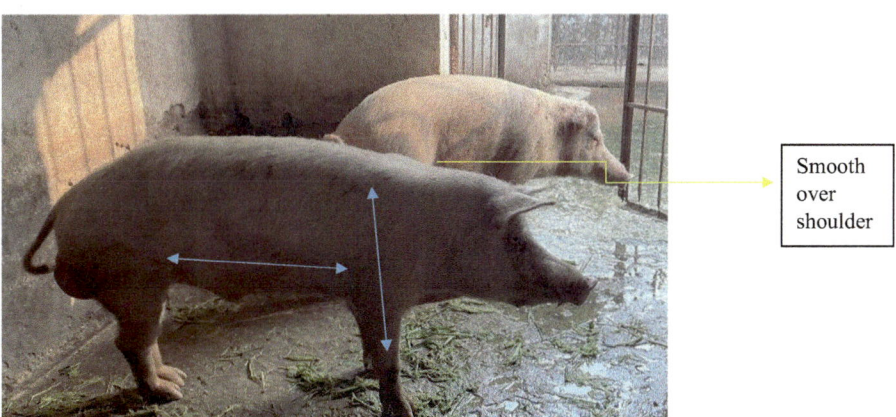

Fig. 7.1 Good body length, correct slope and set of shoulder of a pig

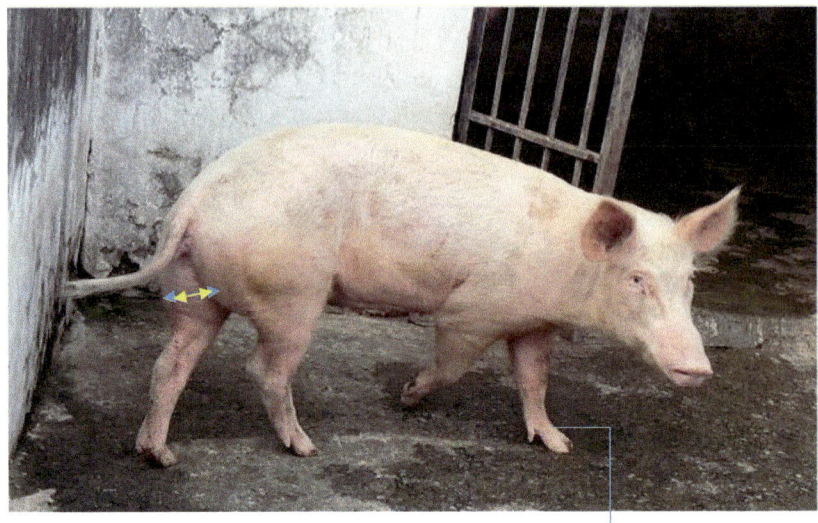

Good feet that are squarely set

Fig. 7.2 Good width and muscle expression of a pig

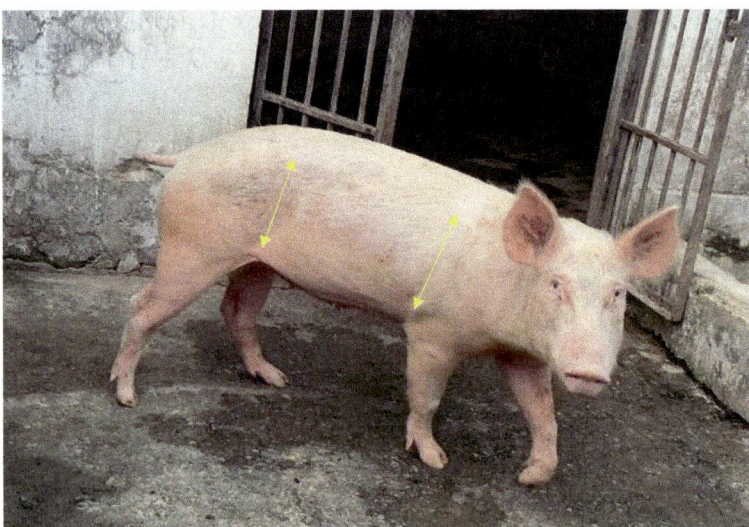

Fig. 7.3 Uniform depth of a pig body

more muscles the pig can build in. The body depth must be uniform from the front to the rear flank, i.e. straight bottom line and top line are good. The length of the pig is measured from the front flank to the hind flank, and this must be the case proportional to the rest of the pig.

(d) **Degree of leanness:** A quality pig has a lot of fat in the belly, cheeks, and back, so there should not be excess fat in pig. To assess thinness, look at the cheek, the shoulder, the flanks, the ham loin joint, and the tail head. The ideal pork cheek extends tightly into the shoulders; shoulder blade moves as pig walks. An indentation is created at the intersection with the ham loin, and there is no longer any loose skin at the starting point of the path. The only fat that counts is intramuscular or marbling. Excess fat will just be like that it is cut off at the butcher shop and adds nothing to the meat you get from pig (Fernandes et al. 2020).

(e) **Structure:** To assess the pig's structure and health, start from the ground up (Pezzuolo et al. 2018). They also reported that all toes of pigs should be pointing forward, ankles and knees straight. The legs should be cleanly boned and the cheeks should be at an angle of about 45 degrees from the floor to ensure structural soundness (Fan et al. 2011). All this allows the pig to move comfortably and smoothly. The top line should be fairly straight, but the rump should slope downward slightly from the end of the loin to the head of the tail. The slanted rump allows the pig farrow easily and take big steps with their hind legs.

(f) **Underline quality:** The pig's bottom line is its belly line. Gilts should have good underlines to raise huge litter, teat accessibility, teat numbers, teat size, and teat placement is important for proper function. About 6–7 teats in each side indicates good underline, teats spaced 2.5–3 inches apart. This follows good and maximum mammary tissue development to feed the piglet and have enough space for the piglet to nurse (Zhuang et al. 2020; Bovo et al. 2021).

7.8 Monitoring and Compliance in the Pig Exhibition

The success and credibility of pig exhibitions depend largely on effective and stringent monitoring compliance measures (Woods et al. 2019). Such events, which combine tradition with contemporary practices, require a structured approach to ensure animal welfare, maintain standards, and promote an environment of integrity. The crucial aspects of monitoring and compliance in pig exhibition are listed below:

7.8.1 Pre-event Inspections and Health Checks

Before starting a pig exhibition, thorough preliminary checks are essential. These inspections serve as a preliminary assessment to ensure that all participating pigs meet the requirements required health standards. Veterinarians carefully examine each animal and check for possible disease signs of illness, ensuring current vaccinations and confirming general well-being. Such proactive measures not only protect the health of the pigs on display, but also prevent potential risks of disease outbreaks that could endanger larger livestock populations (Andreoletti et al. 2011).

7.8.2 On-Site Monitoring During the Exhibition

Once the exhibition is running, ongoing on-site monitoring is of paramount importance (Alhammadi et al. 2023). The educated staff, which often includes veterinarians and agricultural experts, maintains a vigilant presence. Observe animal behaviour, ensure compliance with exhibition protocols, and respond to emergencies concern immediately. Real-time monitoring enables immediate interventions, be it in the health sector anomalies, ensuring proper hygiene or mitigating unforeseen challenges to ensure safety smooth running of the exhibition (Li et al. 2023).

7.8.3 Compliance Enforcement and Penalties for Violations

Enforcement of compliance is non-negotiable. Any deviations from the prescribed rules and standards require immediate attention. The authorities implement a structured enforcement mechanism that this may include warnings, fines or exclusion of non-compliant participants from future exhibitions. Such penalties emphasize the seriousness of compliance and serve as a deterrent by fostering a culture of compliance responsibility and accountability of exhibitors (Busch et al. 2021; Gigot et al. 2024).

7.8.4 Collaboration with Veterinary Professionals and Experts

Collaboration is the backbone of effective monitoring and compliance. Establish robustness partnerships with veterinary experts and subject matter experts enrich monitoring procedure. These collaborations facilitate knowledge sharing, ensure adherence to best practices and offer exhibitors insights into advances in pig farming. In addition, veterinary medicine experts provide invaluable insight during inspections, advice on health protocols and contribute to the refinement of exhibition standards based on evolving scientific understanding (Beltran-Alcrudo et al. 2019).

7.9 Conclusion

The pig exhibition is not just a mere presentation of the livestock, but also a testimony of complicated balance between animal welfare, agricultural practices and regulatory frameworks. Pig exhibitions are not just events for farmers and enthusiasts to display their prized animals; they serve as platforms to showcase breeding successes and promote good practices in pig management and facilitating the exchange of knowledge between those involved. However, with such gatherings come with challenges. Ensuring the health, safety, and welfare of pigs amid there is a lot of hustle and bustle at the exhibitions. This includes compliance with hygienic conditions, preventing disease outbreaks and minimizing stressors that could affect animal health. In order to overcome these challenges, regulatory frameworks are indispensable. The rules leading pig exhibitions are designed with great attention to detail and cover animal health certifications for transportation. These rules are not just bureaucratic in nature. However, these requirements are tailored to ensure animal welfare, public health and uphold the integrity of agricultural practices. In addition, they offer exhibitors a structured framework, ensuring consistent exhibition standards across different platforms. Central to this regulation the context is the need for careful monitoring. Without consistent supervision, most of them do comprehensive regulations run the risk of becoming ineffective. Monitoring ensures that exhibitors comply established guidelines and promotes an environment in which excellence in pig management harmonizes seamlessly with ethical practices. The convergence of strict rules and careful monitoring are essential for a successful pig exhibition. This synergy not only increases the standards of exhibitions, but also underlines the collective obligation to preserve what is essential pig farming, taking modern advances into account.

References

Alhammadi M, Aliya S, Umapathi R et al (2023) A simultaneous qualitative and quantitative lateral flow immunoassay for on-site and rapid detection of streptomycin in pig blood serum and urine. Microchem J 195:109427. https://doi.org/10.1016/j.microc.2023.109427

Andreoletti O, Budka H, Buncic S et al (2011) Scientific opinion on the public health hazards to be covered by inspection of meat (swine). EFSA J 9:2351. https://doi.org/10.2903/j.efsa.2011.2351

Bathelt H, Spigel B (2012) The spatial economy of North American trade fairs. Can Geogr 56:18–38. https://doi.org/10.1111/j.1541-0064.2011.00396.x

Beltran-Alcrudo D, Falco JR, Raizman E, Dietze K (2019) Transboundary spread of pig diseases: the role of international trade and travel. BMC Vet Res 15:64. https://doi.org/10.1186/s12917-019-1800-5

Bosi P, Russo V (2004) The production of the heavy pig for high quality processed products. Ital J Anim Sci 3:309–321. https://doi.org/10.4081/ijas.2004.309

Bovo S, Ballan M, Schiavo G et al (2021) Single-marker and haplotype-based genome-wide association studies for the number of teats in two heavy pig breeds. Anim Genet 52:440–450. https://doi.org/10.1111/age.13095

Bowman AS, Nolting JM, Workman JD et al (2016) The inability to screen exhibition swine for influenza A virus using body temperature. Zoonoses Public Health 63:34–39. https://doi.org/10.1111/zph.12201

Brophy JE, Park J, Bowman AS et al (2024) Understanding if the reward is worth the influenza risk: the true cost of showing pigs. Prev Vet Med 222:106083. https://doi.org/10.1016/j.prevetmed.2023.106083

Bünger A, Schiller D (2022) Identification and characterization of potential change agents among agri-food producers: regime, niche and hybrid actors. Sustain Sci 17:2187–2201. https://doi.org/10.1007/s11625-022-01184-1

Busch F, Haumont C, Penrith ML et al (2021) Evidence-based African swine fever policies: do we address virus and host adequately? Front Vet Sci 8:637487. https://doi.org/10.3389/fvets.2021.637487

Cha E, Toribio J-ALML, Thomson PC, Holyoake PK (2009) Biosecurity practices and the potential for exhibited pigs to consume swill at agricultural shows in Australia. Prev Vet Med 91:122–129. https://doi.org/10.1016/j.prevetmed.2009.05.010

Clark EM, Merrill SC, Trinity L et al (2020) Using experimental gaming simulations to elicit risk mitigation behavioral strategies for agricultural disease management. PLoS One 15:e0228983. https://doi.org/10.1371/journal.pone.0228983

Clark EM, Merrill SC, Trinity L et al (2021) Emulating agricultural disease management: comparing risk preferences between industry professionals and online participants using experimental gaming simulations and paired lottery choice surveys. Front Vet Sci 7:556668. https://doi.org/10.3389/fvets.2020.556668

Cochran HJ, Bosco-Lauth AM, Garry FB et al (2023) African swine fever: a review of current disease management strategies and risks associated with exhibition swine in the United States. Animals 13:3713. https://doi.org/10.3390/ani13233713

DAHD (2024) Department of Animal Husbandry and Dairying, Ministry of Fisheries, Animal Husbandry and Dairying, Government of India. https://dahd.nic.in/

Darian-Smith K, Wills S (2001) From queen of agriculture to miss showgirl: embodying rurality in twentieth-century Australia. J Aust Stud 25:17–31. https://doi.org/10.1080/14443050109387717

Dressler AE, Scheibel RP, Wardyn S et al (2012) Prevalence, antibiotic resistance and molecular characterisation of Staphylococcus aureus in pigs at agricultural fairs in the USA. Vet Rec 170:495. https://doi.org/10.1136/vr.100570

Elzen B, Geels FW, Leeuwis C, van Mierlo B (2011) Normative contestation in transitions 'in the making': animal welfare concerns and system innovation in pig husbandry. Res Policy 40:263–275. https://doi.org/10.1016/j.respol.2010.09.018

Fan B, Onteru SK, Du ZQ et al (2011) Genome-wide association study identifies loci for body composition and structural soundness traits in pigs. PLoS One 6:e14726. https://doi.org/10.1371/journal.pone.0014726

Felicioni F, Pereira AD, Caldeira-Brant AL et al (2020) Postnatal development of skeletal muscle in pigs with intrauterine growth restriction: morphofunctional phenotype and molecular mechanisms. J Anat 236:840–853. https://doi.org/10.1111/joa.13152

Fernandes AFA, Dórea JRR, Valente BD et al (2020) Comparison of data analytics strategies in computer vision systems to predict pig body composition traits from 3D images. J Anim Sci 98:skaa250. https://doi.org/10.1093/jas/skaa250

Gangil D, Singh A, Verma HK, Kansal SK (2019) Perception of the farmers regarding utility of Kisan Mela. Indian J Ext Educ 55:172–175

Gigot C, Lowman A, Ceryes CA et al (2024) Industrial hog operation workers' perspectives on occupational exposure to zoonotic pathogens: a qualitative pilot study in North Carolina, USA. New Solut 33:209–219. https://doi.org/10.1177/10482911231217055

Holloway L, Morris C (2014) Viewing animal bodies: truths, practical aesthetics and ethical considerability in UK livestock breeding. Soc Cult Geogr 15:1–22. https://doi.org/10.1080/14649365.2013.851264

Hunter CL, Millar J, Toribio J-A LML (2022) More than meat: the role of pigs in Timorese culture and the household economy. Int J Agric Sustain 20:184–198. https://doi.org/10.1080/14735903.2021.1923285

Jayaraman B, Nyachoti CM (2017) Husbandry practices and gut health outcomes in weaned piglets: a review. Anim Nutr 3:205–211. https://doi.org/10.1016/j.aninu.2017.06.002

Killian ML, Swenson SL, Vincent AL et al (2013) Simultaneous infection of pigs and people with triple-reassortant swine influenza virus H1N1 at a U.S. county fair. Zoonoses Public Health 60:196–201. https://doi.org/10.1111/j.1863-2378.2012.01508.x

Kumar Rajak S et al (2020) Prospects of animal husbandry in doubling farmer's income by 2022: a review. J Entomol Zool Stud 8:1696–1700

Kyriazakis I (2006) Whittemore's science and practice of pig production. Wiley, Oxford

Lauterbach SE, Nelson SW, Robinson ME et al (2019) Assessing exhibition swine as potential disseminators of infectious disease through the detection of five respiratory pathogens at agricultural exhibitions. Vet Res 50:63. https://doi.org/10.1186/s13567-019-0684-5

Li Y, Fu C, Yang H et al (2023) Design of a closed piggery environmental monitoring and control system based on a track inspection. Robot Agric 13:1501. https://doi.org/10.3390/agriculture13081501

Manu H, Baidoo SK (2020) Nutrition and feeding of swine. In: Animal agriculture. Elsevier, Boca Raton, pp 299–313

Murray AC (2020) The evaluation of muscle quality. In: Quality and grading of carcasses of meat animals. CRC Press, Boca Raton, pp 83–107. https://doi.org/10.1201/9781003068297-4

National Pork Board (2016) A champions guide to youth swine exhibition. In: Michigan State University. https://www.canr.msu.edu/resources/a_champions_guide_to_youth_swine_exhibition_biosecurity_your_pig_project

Nielsen SS, Alvarez J, Bicout DJ et al (2022) Welfare of pigs on farm. EFSA J 20:e07421. https://doi.org/10.2903/j.efsa.2022.7421

Pezzuolo A, Milani V, Zhu DH et al (2018) On-barn pig weight estimation based on body measurements by structure-from-motion (SfM). Sensors (Switzerland) 18:3603. https://doi.org/10.3390/s18113603

Pomar C, Andretta I, Remus A (2021) Feeding strategies to reduce nutrient losses and improve the sustainability of growing pigs. Front Vet Sci 8:742220. https://doi.org/10.3389/fvets.2021.742220

Ressia S, Strachan G, Rogers M et al (2022) Farm businesswomen's aspirations for leadership: a case study of the agricultural sector in Queensland, Australia. Front Sustain Food Syst 6:838073. https://doi.org/10.3389/fsufs.2022.838073

Reynnells R (2012) National extension workshop Washington update. J Appl Poult Res 21:193–200. https://doi.org/10.3382/japr.2011-00480

Sarjahani A, Harmon AH (2015) Considerations for exploring livestock as a nutrition intervention in the rural United States. J Hunger Environ Nutr 10:390–408. https://doi.org/10.1080/19320248.2014.929541

Schneider JD, Tokach MD, Goodband RD et al (2006) Determining the effect of restricted feed intake on developing pigs weighing between 150 and 250 lb, fed two or six times daily. Kansas Agric Exp Stn Res Rep:34–38. https://doi.org/10.4148/2378-5977.6984

Schweer WP, Schwartz K, Burrough ER et al (2016) The effect of porcine reproductive and respiratory syndrome virus and porcine epidemic diarrhea virus challenge on growing pigs I: growth performance and digestibility. J Anim Sci 94:514–522. https://doi.org/10.2527/jas.2015-9834

Shurson GC, Urriola PE, Schroeder DC (2023) Biosecurity and mitigation strategies to control swine viruses in feed ingredients and complete feeds. Animals 13:2375. https://doi.org/10.3390/ani13142375

Sun H, Palaoag T, Quan Q (2022) Design of automatic monitoring and control system for livestock and poultry house environment based on Internet of Things robot. In: 2022 4th Asia Pacific information technology conference. ACM, New York, pp 224–230

Święch E (2017) Alternative prediction methods of protein and energy evaluation of pig feeds. J Anim Sci Biotechnol 8:39. https://doi.org/10.1186/s40104-017-0171-7

Tummaruk P, Tantasuparuk W, Techakumphu M, Kunavongkrit A (2007) Age, body weight and backfat thickness at first observed oestrus in crossbred Landrace×Yorkshire gilts, seasonal variations and their influence on subsequence reproductive performance. Anim Reprod Sci 99:167–181. https://doi.org/10.1016/j.anireprosci.2006.05.004

Vincent AL, Swenson SL, Lager KM et al (2009) Characterization of an influenza A virus isolated from pigs during an outbreak of respiratory disease in swine and people during a county fair in the United States. Vet Microbiol 137:51–59. https://doi.org/10.1016/j.vetmic.2009.01.003

von Borell E, Bonneau M, Holinger M et al (2020) Welfare aspects of raising entire male pigs and immunocastrates. Animals 10:1–12. https://doi.org/10.3390/ani10112140

Wemelsfelder F, Nevison I, Lawrence AB (2009) The effect of perceived environmental background on qualitative assessments of pig behaviour. Anim Behav 78:477–484. https://doi.org/10.1016/j.anbehav.2009.06.005

Woods AL, Tynes VV, Mozzachio K (2019) Special considerations for show and pet pigs. In: Diseases of swine. Wiley, Hoboken, pp 211–220

Zhao D, Yang B, Yuan X et al (2021) Advanced research in porcine reproductive and respiratory syndrome virus co-infection with other pathogens in swine. Front Vet Sci 8:699561. https://doi.org/10.3389/fvets.2021.699561

Zhuang Z, Ding R, Peng L et al (2020) Genome-wide association analyses identify known and novel loci for teat number in Duroc pigs using single-locus and multi-locus models. BMC Genomics 21:344. https://doi.org/10.1186/s12864-020-6742-6

Impact of Mycotoxins on Pig Production

Amitava Roy and Tanmoy Rana

Abstract

Feed for agricultural animals frequently contains mycotoxins. Mycotoxins have a great potential to harm sows and gilts. This chapter provides an overview of the primary mycotoxins found in pig feed that impair sow fertility and reproduction. Depending on the type of mycotoxin, the amount and length of exposure, the animal's health and condition at the time of exposure, and other factors, eating feed contaminated with these mycotoxins can result in a wide range of symptoms. It is known that there are two kinds of fungi: storage fungi and field fungi. Toxins produced by field fungi such as Fusarium species, Aspergillus species, and Claviceps species can impair reproductive function. If the humidity is too high while being stored, storage fungus will grow. In everyday life, mycotoxicosis symptoms might manifest at levels of toxin below the limit of detection. The consequences of mycotoxins are becoming more and better understood. Due to the existence of hot spots in feed and/or feedstuffs, mycotoxins may still be present in feedstuffs even in the case of negative analytical results. Clinical signs can be fairly prominent, which makes the practitioner's diagnosis relatively simple. However, in many situations, the symptoms are ambiguous or non-existent at all at the herd level on a regular basis. Whenever there is even the slightest suspicion that a mycotoxicosis issue may be leading to reproductive failure in breeding pigs, the practitioner is in the forefront of educating all parties involved. Only by taking the required precautionary steps and activities will all parties concerned in

---

A. Roy (✉)
Department of Livestock Farm Complex, West Bengal University of Animal & Fishery Sciences, Kolkata, India

T. Rana
Department of Veterinary Clinical Complex, West Bengal University of Animal and Fishery Sciences, Kolkata, West Bengal, India

© The Author(s), under exclusive license to Springer Nature Singapore Pte Ltd. 2024
T. Rana, B. Soto-Blanco (eds.), *Good Practices and Principles in Pig Farming*, Livestock Diseases and Management,
https://doi.org/10.1007/978-981-97-4665-1_8

the health of the pig herd be able to fix the issues. This chapter mostly discusses aflatoxins, ergot alkaloids, trichothecenes, and zearalenone as the primary toxins that cause reproductive failure.

Keywords

Mycotoxins · Reproduction · T-2 · Aflatoxins · Zearalenone · Ergot · Storage fungi

8.1 Introduction

Toxic by-products of naturally occurring metabolic processes in fungi are called mycotoxins. The four primary genera of fungus that produce mycotoxin are Fusarium spp., Penicillium spp., Claviceps spp., and Aspergillus spp. When plants and animals are being produced, mould can grow at different stages. Before the crop is actually harvested, while it is still in the field, mycotoxins might infiltrate the seeds. Alternatively, mould growth can happen when the grain is being stored at the feed mill or on the farm. Because of this, ingredients may already contain significant amounts of mycotoxins when they are delivered to feed mills or farms. Therefore, it can be quite challenging to stop mycotoxin incidence in feed ingredients. During feed processing, mould can also develop, particularly if the mixer raises the feed's temperature and humidity. Finally, inadequate cleaning of silos, transport systems, and feeds at the farm level may also lead to the growth of mould and the formation of mycotoxin. Various feed ingredients, including maize, wheat, barley, millet, peanuts, peas, and oily feedstuffs, can become contaminated by the fungi. The humidity of the substrate (10–20%), the relative humidity ($\geq$70%), the temperature (0–50 °C, depending on the kind of fungus), and the availability of oxygen all contribute to the synthesis of mycotoxins. Mycotoxicosis or other toxic consequences can be caused by mycotoxins. The degree and kind of mycotoxin contamination, as well as a number of other variables like animal species, sex, the environment, nutritional state, and other toxic substances, all affect the symptoms. They are not, however, transmissible from one animal to another, and tainted feed is primarily to blame. Because the signs of mycotoxicosis in animals can range from particular to non-specific, such as immune suppression, diarrhoea, bleeding, or decreased performance, diagnosing the illness is frequently quite challenging. In general, there is a regional pattern to the presence of mycotoxins in food and animal feed. For example, Aspergillus species thrive in tropical and subtropical climates, while Fusarium and Penicillium species are better suited to the climates of North America and Europe. On the other hand, mycotoxin issues arise globally as a result of the international trade in food and feed. Mycotoxins contaminate 25% of the world's crop, according to Lawlor and Lynch (2001). Trichothecenes, zearalenone, ochratoxins, aflatoxins, and fumonisins are the five major types of mycotoxins from the perspective of animal production. This chapter discusses how mycotoxins affect pig reproductive.

8.2 Fungi Essential to Pig Procreation

Reproductive mycotoxicoses are characterised by a decrease in reproductive performance and reproductive failure caused by mycotoxins. Fungus can be classified into two categories: pathogenic fungus, also known as field fungi, which cause plant diseases before harvest, and saprophytic fungi, also known as storage fungi, which can only survive on dead organic matter after harvest. In both situations, the fungi need particular environmental factors to flourish. Depending on the kind of fungus, these circumstances could include high temperatures, oxygen, and relatively high humidity (Desjardins 2006). Aspergillus species, Fusarium species, and Claviceps species are the three main genera of mycotoxin-producing pathogenic fungus that are crucial for sow reproduction. The mycotoxins that these three genera produce are harmful to pigs, not the fungi themselves. It is not always harmful when toxins-producing fungus are found in animal feed or raw materials. Multiple fungi can infect a feedstock, and those fungi can produce multiple mycotoxins. As a result, it frequently happens that many mycotoxins manifest concurrently in feeds (Koshinsky and Khachatourians 1992). Because of the additive or synergistic interaction between the two, this combination may have more negative consequences than a single mycotoxin.

8.3 Development of Mycotoxin

Only specific moulds and circumstances can cause mycotoxins. Consequently, the existence of mould in feedstuffs or feeds does not always indicate the presence of mycotoxins. The two main variables affecting the growth of mould and the production of mycotoxin are temperature and moisture. Easy-to-get starch from grains, moisture, air, and the right temperature—typically between 54 and 77 °F—are all necessary for mould formation. Under such circumstances, mycotoxin formation is possible, but stressors like drought, high temperatures, excessive watering, nutrient shortage, insect damage, and harvest damage are usually also required to predispose the formation of mycotoxins (Rodrigues and Naehrer 2012). Moulds that produce mycotoxin are divided into two groups: field moulds and storage moulds (Osweiler and Ensley 2012). Prior to harvest, field moulds thrive in grains and need high relative humidity (above 70%) and grain wetness (above 22%). Fusarium species are the most worrisome field moulds because they can create fumonisin, zearalenone, and vomitoxin. After harvesting and while grains and feeds are being stored, storage moulds develop in the grains. These moulds can even grow in grains with 12–18% moisture content and do not require high humidity. Aflatoxin and ochratoxin are produced by Aspergillus and Penicillium species, which are storage moulds. Under some circumstances, field moulds continue to proliferate during storage whereas storage moulds continue to grow in grains even before harvest. This is frequently the result of the mould *Aspergillus flavus*, which generates aflatoxin. Additionally, grains and feeds may have many moulds and multiple mycotoxins present at the same time.

8.4 Contamination with Mycotoxin

Contamination with mycotoxin happens everywhere. There are certain areas that are more likely to be contaminated by mycotoxin. For instance, tropical and subtropical areas often have higher storage mould prevalence (Aspergillus and Penicillium species) while temperate regions tend to have higher field mould prevalence (Fusarium species). Mycotoxins have a broad impact on grains from a region where field mould growth circumstances are favourable. In contrast, grains are not equally impacted by mycotoxins and distribution is more varied both within and within storage bins when conditions are favourable for storage moulds (Jacela et al. 2010). Grain contamination, particularly in corn, sorghum, wheat, and barley, is a common occurrence. According to Hendel et al. (2017), corn is the grain that is most heavily polluted. Furthermore, Rodrigues and Naehrer (2012) found that mycotoxins are frequently concentrated in grain co-products such maize distillers dried grains with solubles (DDGS). Most of the starch in maize is removed during the fermentation process that creates DDGS. If the corn is infected, mycotoxins remain untouched by the process and can concentrate up to three times their original concentration (Jacela et al. 2010).

8.5 Pig Reproductive Health Depends on Mycotoxins

The primary mycotoxins found in pig reproduction are trichothecenes, which are represented by T-2, ergot alkaloids, and zearalenone, which is a significant toxin. These poisons are linked to numerous diseases and cases of infertility and can be found in feed and feed components.

8.6 Zearalenone

The primary mycotoxins found in pig reproduction are trichothecenes, which are represented by T-2, ergot alkaloids, and zearalenone, which is a significant toxin. These poisons are linked to numerous diseases and cases of infertility and can be found in feed and feed components. Mycotoxin zearalenone is generated by Fusarium species, primarily *F. graminearum* and *F. culmorum*. The most commonly impacted crops are maize and wheat, and this mycotoxin is produced more readily when there is a high relative humidity during storage (Desjardins 2006). With its resorcylic acid lactone structure, zearalenone can penetrate cell membranes and attach itself to cytosolic 17β-estradiol (E2) receptors to form a complex known as ZEA-E2R. After being transferred into the cell nucleus, this complex (ZEA-E2R) attaches to particular nuclear E2 receptors, activating the gene that codes for mRNA synthesis, which is typically produced by E2. These effects resemble those of oestrogen and stimulate anabolism and reproduction. In addition to interacting with both kinds of oestrogen receptors, zea is a substrate for the enzyme hydrosteroid dehydrogenases, which break it down into alpha- and beta-zearalenol, two stereoisomeric metabolites. While glucuronidation capacity inactivates ZEA, alpha-hydroxylation increases the efficacy of oestrogen and explains the species-specific vulnerability to ZEA intoxications. Pigs are more susceptible to ZEA due to their lower glucuronidation capacity as compared to other species (Fink-Gremmels and Malekinejad 2007). Zearalenone has a hyperestrogenic effect and pigs are particularly sensitive to it (Andretta et al. 2008). Anoestrus, abortion, increased foetal and embryonic death, failure of PGF2α induction programmes, and a higher rate of stillbirth and splay-legged piglets are the most common pathological outcomes (Alexopoulos 2001). According to Edwards et al. (1987a), gilts are more sensitive than sows.

8.7 Oestrus Cycle

8.7.1 Gilts

Experimental feeding of contaminated or zearalenone feed at relatively low doses (1.5 to 2 ppm) causes atrophic ovaries, increased uterine mass, and vaginal and vulvar wall swelling in gilts, but does not cause a standing reflex (Andretta et al. 2008). Symptoms start to show up 3–7 days after the first dose and go away 14 days after the conditioned feed is stopped (Kordic et al. 1992). When injecting 2 mg of estradiol benzoate or estradiol-17 β cyclopentyl proprionate (Young et al. 1990), similar effects are observed (Flowers et al. 1987). While adult cyclic and lactating sows do not respond similarly to these concentrations, gilts do (Edwards et al. 1987b). Significantly larger dosages (64 ppm) are needed in sows to produce comparable symptoms (Long et al. 1982). Zearalenone concentrations of 3 ppm can cause anoestrus in gilts (Young and King 1986). If given to gilts at a level of 2 ppm for 45–90 days, zearalenone can likewise cause early puberty in them (at 70 days of age) (Rainy et al. 1990). But without ovulation, the majority of these gilts' initial heats are sterile. According to recent studies, the intrinsic quality of the oocytes taken from gilts contaminated with modest quantities of zearalenone (0.235–0.358 ppm) was dramatically decreased when diet was given to them (Alm et al. 2006). According to Ranzenigo et al. (2008), the degree of contamination affects granulosa cells, steroidogenesis, and gene expression in a dose-dependent manner.

8.7.2 Embryos

Numerous research have looked into the potential effects of zearalenone during pregnancy (see, for example, Christensen et al. 1972). Smaller litters are produced by pregnant gilts and sows who are fed contaminated feed (>2.8–3.0 ppm ZEA), particularly in the early stages of pregnancy. The appropriate secretory response of the endometrium to progesterone (P4) during embryo implantation on Days 11–12 following breeding is disrupted by premature estrogenic stimulation by ZEA. Severe

ZEA exposure (>25 ppm ZEA in feed) will result in symptoms like as stillbirth and neonatal mortality. A common occurrence in feed with 2.8–3.0 ppm ZEA is "mummification." Towards the end of gestation, splay-leg and neonatal oestrogen syndrome in newborn pigs are caused by PGF2α-induced programmed farrowing induction that can be challenging to implement. This causes the vulva to swell dramatically in one-day-old female piglets. No effect on ovulation or nidated foetuses has been demonstrated by the experimental administration of feed containing 10 ppm zearalenone (Rainy et al. 1990).

In a similar vein, neither the number of piglets born nor their survival rate was affected by the experimental addition of 9 ppm zearalenone to the feed throughout the whole gestation period (Young and King 1986). Zearalenone treatment appears to be particularly crucial between Days 7 and 10 of pregnancy, as there is a larger rate of embryonic death during this time than there is before or after (Long and Diekman 1986). When gilts were given 1 ppm experimentally between Days 7 and 10 of pregnancy, there was an increasing deterioration of embryos beginning on Day 9. By Day 13, the surviving blastocysts were degenerating, the researchers noticed (Long et al. 1992). Zearalenone is thought to affect the endometrium's secretory process, altering the intrauterine environment during the early stages of pregnancy, even if the many studies did not identify a clear cause for the early embryonic demise (Etienne and Jemmali 1982). Conversely, intermediate amounts of zearalenone pollution (up to 60 ppm) result in smaller litters and less active piglets, while higher levels (64 ppm) cause the death of the entire litter (Long et al. 1982). Long et al. (1982) revealed a dose-related effect by demonstrating an increase in the number of tiny litters (1–3 piglets) with incremental zearalenone contamination rates (0, 7, 38, and 64 ppm). Young et al. (1981) similarly showed an inverted linear association between the amount of contamination and litter size. Zearalenone's effect on litter size can be attributed to both the piglets' embryonic and foetal deaths as well as its detrimental effect on fertilisation. Most likely, this is because of how negatively it affects the luteinizing effect (Obremski et al. 2003).

8.7.3 Foetuses

The foetuses' weight is decreased when sows receive an experimental 4 ppm zearalenone meal throughout their whole gestation (Etienne and Jemmali 1982). Additionally, it causes the weight differences within piglets in the same litter to increase. Young and King (1986) discovered a negative correlation between the average birth weight of the piglets and the plasma level of zearalenone. This could account for the rise in piglets delivered with splay-leg from sows fed more than 4 ppm zearalenone, given the range in birth weight. The existence of foetal tissue remnants in the uterus on Day 40 following fertilisation was used to establish foetal death (Long and Diekman 1984).

8.7.4 Newborn Piglets

What happens to nursing sows when they are fed feed tainted with zearalenone is not well understood. It has been demonstrated that sows fed feed containing 4.8 ppm of zearalenone during gestation and lactation experienced greater piglet mortality during the first 2 weeks following birth. Zearalenone or its metabolites (α- and β-zearalenone) may have harmful effects on piglets if they are exposed to the sow's milk (Palyusik et al. 1980).

8.7.5 Sows

Zearalenone contamination in feed at 5–10 ppm results in a protracted cycle or even anoestrus in cyclic sows after weaning (Meyer et al. 2000). Young et al. (1990) demonstrated a linear relationship between the duration of anoestrus in days and the zearalenone level in ppm. If gilts are fed in the same amounts from Day 5 to Day 20 of the cycle, this effect may also be observed in them. The elevated progesterone level keeps the luteal bodies active. Thirty days after the contaminated feed is removed, the persistent corpora lutea spontaneously dissolve (Edwards et al. 1987b). Zearalenone is assumed to directly affect the ovaries because it has no effect on gonadotropic hormones (Flowers et al. 1987). Nevertheless, Diekman et al. (1986) discovered that extremely high dosages (1 mg/kg body weight) also suppress the release of FSH and LH in an ovariectomised gilt.

8.7.6 Boars

Over the course of 6–15 weeks, young, immature boars fed contaminated feed containing zearalenone (up to 600 ppm) had testicular weights that were up to 30% lower than those of control boars (Young and King 1983). The vesicular glands and the epididymis are also smaller, even after feeding with 100 ppm of zearalenone. Spermatogenesis has been shown to be momentarily inhibited; however, this may be overcome by removing the contaminated feed. After administering 40 ppm zearalenone, a lower libido was noted, which was correlated with a lower plasma testosterone concentration (Berger et al. 1981). However, young boars (those younger than 38 weeks of age) were the only ones exhibiting this impact. The same results—a decline in testosterone and libido—were observed in a different study (Ruhr et al. 1983) with a dosage range of 20–200 ppm given to older boars over the course of eight weeks. It was shown that the amounts employed decreased the capacity of the boar spermatozoa to adhere to the zona pellucida in a recent study employing prolonged boar semen with different doses of ZEA and alpha-zearalenol ranging between 40, 60, and 80 µg/ml (Tsakmakidis et al. 2007).

8.7.7 Trichothecenes

Fusarium spp. generate trichothecenes, such as T-2 and deoxynivalenol (DON). The most significant Trichothecene effecting the reproductive system is T-2.

8.7.8 Ergot Aalkaloids

Claviceps fusiformis, *Claviceps paspalli*, and *Claviceps purpurea* all generate ergot alkaloids. Rye, wheat, and barley are the primary hosts of these pathogenic fungus. The fungus develops a sclerotium within the grain ear. Because the sclerotia are eliminated by modern cleaning and storage methods, ergot feed contamination has become extremely rare (Diekman and Green 1992). Clinical indicators of ergot alkaloids include oxytocin-resistant agalactia, small litters, premature farrowing, mummification, recurrent oestrous, metritis, and mastitis. Ergot alkaloids have an impact on reproductive performance (Barnikol et al. 1982). Prolactin inhibition is the reason for the impact on milk production (Kopinski et al. 2007). In 50% of sows, a concentration of 0.3% sclerotia in the lactation feed is enough to induce agalactia. Within the first 8 days of life, afflicted sows' newborn piglets experience diarrhoea. Lameness is a common symptom in some sows and gilts, especially in the hindquarters, and necrosis on the tail, ears, and hooves is frequently observed. The onset of clinical symptoms occurs within a few weeks, and inclement weather makes them worse (Osweiler et al. 1990). Clinical indicators quickly go down when contaminated feed is simply stopped. Three to seven days after stopping to feed the contaminated feed, milk output returns to normal.

In addition to its reproductive side effects, ergot intoxication can lead to endothelial damage and vasoconstriction, which can result in ischemia and dry gangrene, particularly in the hooves, tail, and ears of unweaned piglets. Pigs affected exhibit decreased feed intake, elevated heart and respiratory frequencies, and wasting. The primary sign of fattening pigs is a decrease in ADG, which is already noticeable at 0.1% ergot in the feed. Elevated concentrations result in a greater loss of ADG and more feed waste.

8.7.9 Deoxynivalenol (DON)

When grains are stored poorly, causing a high humidity of 20–22%, *Fusarium roseum* or *Fusarium graminearum* are the principal producers of deoxynivalenol (DON), also known as vomitoxin. Reduced feed intake results from DON-contaminated feed, and vomiting may occur from high DON levels (Diekman and Green 1992). Studies revealed that fusaric acid (FA) increased the effects of DON (Smith et al. 1997), resulting in feed rejection and vomiting at DON concentrations as low as 0.14 ppm. More FA at the same DON levels results in a more severe DON intoxication. Cereals are often the source of FA in the feed, although it can also come from other feed ingredients such soy beans. In addition to lowering food intake, feeding contaminated feed containing DON can affect the immune system (cellular and humoural), cause metabolic disturbances in the liver and spleen primarily because of inhibition of RNA, DNA, and protein synthesis (Smith and Az-Llano 2009), and alter reproduction, which will reduce the development of oocytes and embryos (Tiemann and Danicke 2007; Ranzenigo et al. 2008). Compared to pregnant sows, prepuberal gilts respond more sensitively to DON > ZEA feeding. The relationship between DON and reproduction in pigs is more indirect; it is primarily associated with decreased feed intake and subsequent malfunctioning of key organs such as the spleen and liver.

8.7.10 Aflatoxicosis

Aspergillus flavus and *Aspergillus parasiticus* are the primary producers of aflatoxins, which are found in a variety of commonly used feedstuffs (Thieu et al. 2008). Based on their fluorescence during chromatography, four aflatoxins are identified as either green (G1, G2) or blue (B1, B2). B1 is thought to be the most poisonous. The primary source of M1, the fifth metabolite, is the milk produced by animals fed feed contaminated with B1. M1, which is extremely carcinogenic, is also present in the tissues and urine of afflicted animals. Pea-nuts, maize, and cotton seeds are the most susceptible feed ingredients and are utilised in commercially available pig feedstuffs. Pork is quite vulnerable to aflatoxins. By attaching itself to nuclear DNA, aflatoxins stop the synthesis of RNA, enzymes, and other proteins. M1's carcinogenic action is explained by its binding to macromolecules and the endoplasmic reticulum (Booth and McDonald, 1982). Depending on the initial contamination of the feed, varied levels of these metabolites may be found in the sow's milk. Sow's milk contains the aflatoxins B1, G1, and M1 (Silvotti et al. 1997). Piglets exposed to aflatoxins during experimental intoxications displayed damaged macrophages and lymphocytes, indicating a loss of immunological competence. Histology may reveal centrilobular necrotic hepatitis in situations of acute aflatoxicosis. Anorexia, neurological symptoms, and abrupt death are among the clinical manifestations of acute aflatoxicosis (Hale and Wilson 1979).

Aflatoxin residues have a relatively short half-life. The average half-life period for feed concentrations between 355 and 551 µg/kg is 24 hours. Only trace amounts of residues (less than 0.5 µg/kg) were discovered after 48 hours, and after 4 days, none remained (Furtado et al. 1982). A logarithmic relationship was found by Jacobson et al. (1978) between the amount of aflatoxin consumed and the concentration of B1, B2, and M1 residues. It has been established that the liver is the best organ to track aflatoxicosis and measure residue levels. Stubblefield et al. (1991) modified the AOAC thin-layer chromatography (TLC) method to an LC with fluorescence detection in order to quantify these amounts of contamination. With the use of this modified technique, it was possible to detect levels of intoxication in the liver of up to 0.04 ppb for B1 and 0.1 ppb for M1 metabolites.

8.7.11 T-2

Fusarium tricinctum produces T-2, one of the most dangerous mycotoxins found in soybeans, wheat, rye, and corn. The hallmark of T-2 mycotoxicosis, sometimes known as "mouldy corn disease," in pigs is numerous haemorrhages on the serosa of the stomach, oesophagus, and liver (at necropsy). There is a cream-coloured paste on the lining of the oesophagus and the ileum, and blood is present in the intestines and the abdominal cavity (Weaver et al. 1978c). T-2 is a potent immunosuppressant due to its radiomimetic action. Pigs raised on an experimental diet supplemented with 2 or 3 mg/kg of feed show a decrease in MCV and haemoglobin levels in addition to a decrease in red blood cell count. Moreover, there may be a notable decrease in the quantity of T cells. It is a dose-related effect (Rafai et al. 1995). T-2 also has a significant effect on pigs' ability to reproduce. According to one study (Glavits et al. 1983), giving sows contaminated feed at a concentration of 1–2 ppm during the final third of their gestation had an inhibitory impact on the ovaries, resulting in histopathological degeneration and atrophy. Weaver et al. (1978a, b) found that experimental poisoning of sows with T-2 contaminated feed at a level of 12 ppm for 220 days results in repeat breeders and small litters with underweight piglets. T-2 had no effect on the sows' overall health in this investigation, and the piglets showed no signs of lesions. Another study (Vanyi et al. 1991) examined the effects of T-2 on piglets when pregnant sows were experimentally intoxicated with 24 mg of T-2 per day during the last third of gestation. These sows gave birth to piglets that perished shortly after, exhibiting unconsciousness, diarrhoea, and wasting. T-2 metabolites were discovered in both the piglets' stomach contents and the sow's milk. Hypoglycaemia from a decrease in hepatic glycogen may be the cause of the coma.

8.7.12 Diagnoses and Therapies

The mycotoxins discussed in this review of the literature can result in particular symptoms related to reproduction. The age and stage of life of the animal at the time of contamination, in addition to the concentration in the feed and the length of exposure, will determine the reported symptoms. In controlled experimental settings with a single toxin contamination, establishing cause and effect is easier than in field settings where many toxin contaminations may occur. Pigs may reject tainted food, hence it could be required to inject or inoculate the mycotoxin even in experimental settings. Mycotoxin intoxication typically results in reproductive failure in a small number of animals and is not often considered an issue for the herd. The golden rule still applies in any questionable situation: a thorough history and anamnesis of the farm's reproductive management, accounting for handlings, imports and movements of gilts, sows, and/or boars, as well as existing feeding practices and sanitation, should be conducted. The veterinarian will be able to construct a suitable differential diagnosis with the help of this anamnesis. The practitioner must counsel the farmer to implement preventive and farm management control measures if he is clinically confident that mycotoxin poisoning is the root cause of the reproductive failure. As of yet, there is no particular treatment available, except than prevention or stopping the contaminated feed.

8.8 Mycotoxin-Induced Immunomodulation's Effects on Pig Health

8.8.1 Availability to Contagious Illnesses

According to Antonissen et al. (2014), mycotoxins have a wide immunosuppressive effect that may reduce host resistance to infectious illnesses. *Erysipelothrix rhusiopathiae* infections in pigs are more severe when they consume feed contaminated with AF (Cysewski et al. 1978). In a similar vein, eating AF shortened the incubation period and made diarrhoea more severe during an experimental infection with *Brachyspira hyodysenteriae* (Joens et al. 1981). DON intensifies the viral infection when the pig reproductive and respiratory syndrome virus (PRRSV) or porcine circovirus type 2 (PCV2) virus is present. It also intensifies the infection with more tissue lesions generated when the PRRSV and DON are present (Savard et al. 2015b). DON causes a rise in pro-inflammatory cytokine production during a bacterial infection, which intensifies the inflammatory response (Vandenbroucke et al. 2011). On the other hand, a low level of DON may enhance the resistance to some infections due to the rise in circulating IgA. In fact, numerous immune-related genes are rapidly and transiently up-regulated when IgA is present (Pestka et al. 2004). Ingestion of FB1 in pigs can result in intestinal infections and compromise certain intestinal functions (Burel et al. 2013). Consuming feed tainted with FB1 was also linked to a higher risk of developing lung infections and a worsening of pathological alterations caused by bacterial or viral pathogens (Posa et al. 2013). Consuming feed tainted with OTA increases one's vulnerability to infectious diseases that arise naturally. In fact, all piglets fed an OTA-contaminated meal developed salmonellosis on their own. Even after the animals had a vaccination against the disease, eating contaminated feed causes spontaneous infections with Brachyspira hyodysenteriae and *Campylobacter coli* (Stoev et al. 2000). OTA raises the viremia in sera and tissues during a PCV2 infection (Gan et al. 2015). As far as we are aware, there are no data on the impact of ZEN on in vivo analysis on mice that were fed 10 mg/kg ZEA (1.5 mg/kg BW daily) for 2 weeks while being infected with *Listeria monocytogenes*. When compared to control animals, these mice displayed a decreased resistance to Listeria with an increasing trend of the splenic bacterial counts (Pestka et al. 1987).

8.8.2 Chronic Infection Reactivation

It was also looked into how mycotoxin poisoning affected the reactivation of persistent infections. Nevertheless, rats rather than pigs were used in the experiment. The presence of encysted parasites indicates the progression of the *Toxoplasma gondii* infection to a chronic phase in the immunocompetent host. Cyst rupture is possible, but reactivation is inhibited and the infection stays latent. In individuals afflicted with the human immunodeficiency virus and other immunosuppressed animal subjects, rupture is linked to the development of additional cysts and illness. In previously infected mice, Toxoplasma cyst rupture can be accelerated by low and repeated doses of either AFB1 or T-2 toxin (Venturini et al. 1996).

8.8.3 Vaccination Efficacy

Ingestion of mycotoxin can potentially compromise immunization-induced immunity. For instance, AFB1 prevents swine from developing acquired immunity after erysipelas vaccination using *E. rhusiopathiae* bacterin preparation, which is a suspension of dead bacteria (Cysewski et al. 1978). As previously noted, eating feed tainted with AFB1 or T-2 toxin inhibited the vaccination response to ovalbumin, the model antigen, affecting the humoural and cellular responses, respectively (Meissonnier et al. 2008). Pigs' specific antibody response to the Mycoplasma vaccine is reduced when they consume low doses of FB1, another mycotoxin (Taranu et al. 2005). Pigs immunised against Aujesky illness (Suid Herpesvirus 1 [SuHV1]) then exposed to OTA or FB1 showed significantly disrupted humoural immune responses, including a significant drop in antibody levels (Stoev et al. 2012). Pigs receiving an OVA vaccine under a diet tainted with DON or FB1 exhibited a change in their particular immunological response (Grenier et al. 2011).

Similarly, it was demonstrated that giving pigs a diet tainted with DON would significantly reduce virus replication, thereby reducing the effectiveness of the live PRRSV vaccine (Savard et al. 2015a). It should be noted that at mycotoxin dosages that do not change the overall immune response, the vaccination immune response is changed (Meissonnier et al. 2008). Illness outbreaks even in flocks that have received the recommended vaccinations due to a loss in vaccine immunity. These responses have significant implications for animals whose health is dependent on a successful immunisation programme.

8.9 The Issue of co-Contamination with Mycotoxins

The effects of a single mycotoxin on immunity were discussed in the paragraphs above. Nevertheless, animals are frequently exposed to multiple mycotoxins at once and mycotoxins frequently coexist. It is true that many fungi can contaminate raw materials and produce multiple mycotoxins at the same time. Additionally, animals consume a variety of commodities in their diet (Alassane-Kpembi et al. 2016; Rodrigues and Naehrer 2012). According to other research, between 75% and 100% of animal feed samples include multiple mycotoxin contamination (Streit et al. 2012). It is not always possible to forecast the toxicity of mycotoxin combinations solely on the toxicities of the individual ingredients. It can intensify the effects of each mycotoxin in an antagonistic, additive, or synergistic manner.

There are few investigations on the effects of a cocktail of mycotoxins on pig immune response. Numerous pig in vivo studies have looked into the reduction of lymphocyte proliferation. Different types of interactions were found, including additivity (co-exposure to AF and FB at 0.05 and 30 mg/kg feed] for 1 month; synergy (co-exposure to FB and DON at 50 and 4 mg/kg feed] for 28 days) or OTA and T-2 toxin at 2.5 and 8 mg/kg feed) for 30 days (Grenier and Oswald 2011). Over the course of 35 days, animals coexposed to DON and FB (6 and 3 mg/kg of feed) showed antagonistic interactions on levels of specific IgA and cytokine expression, additive interactions on cytokines expression (IL-8; IL-1b, IL-6, and macrophage inflammatory protein 1b), and synergistic interactions on lymphocyte proliferation upon mitogenic stimulation (Grenier et al. 2011). Over the course of 35 days, animals co-exposed to DON and FB (6 and 3 mg/kg of feed) showed antagonistic interactions on levels of specific IgA and additive interactions on specific IgG, lymphocyte proliferation following mitogenic stimulation, and cytokine expression (IL-8, IL-1b, IL-6, and macrophage inflammatory protein 1b).

8.10 Control and Prevention

Before the feed enters the feed mills or animal feeds (e.g. in the field), mycotoxin production can already be occurring. It might not always be possible to prevent the occurrence of mycotoxin contamination in these situations. Thus, the only option is frequently to keep animals from becoming contaminated and from suffering from mycotoxin intoxication. It's important to recognise the effects that mycotoxins have on an animal's health and physiological state as well as the circumstances in which they may be present in feed, such as in dirty feeders or silos. Feed mill quality programmes and routine nutritionist inspections of feed ingredient quality, including potential mycotoxin levels in the feed stuffs, can be very helpful in managing these intoxications and frequently provide a more reliable foundation for the assessment of these toxins than bile and blood tests (Danicke et al. 2008). While it is possible to analyse the amount of mycotoxin in ingredients, doing so is not always simple and necessitates extensive sampling procedures. This is primarily because freshly delivered batches of feedstuffs frequently contain "hot spots," which are localised colonies of various mould families in a batch of feedstuffs. Responsible end users can access an updated information system that displays the contamination levels of feed materials on a European scale through the Rapid Alert system for Food and Feed (RASFF). Visit https://web-gate.ec.europa.eu/rasff-window/portal/ to consult these databases. Inspectors of food or feed examine goods at international crossings or on the market. If, upon sample, the product is found to be non-compliant, it must be reported within the national system. If the relevant authority determines that the problem comes within the purview of the RASFF, they will be notified of this as well. All of the information on the RASFF notification form is collected by templates; since reporting started in 1981, zearalenone, aflatoxins, and Fusarium spp.

have been reported the most frequently (6270, 10 and three times, respectively). The database is dominated by aflatoxins because of their connection to food intended for human consumption.

Numerous new analytical techniques have been created globally, enabling these monitoring programmes to be carried out more effectively. Mycotoxicosis can still be prevented primarily by following the management and hygiene guidelines outlined in the current Good Manufacturing Practices (GMP) manuals and by having a responsible veterinarian oversee the farm. More frequent cleaning and disinfection of the identified critical points at the various levels in the feed mill and/or farm will result from raising awareness of specific critical conditions (such as dust, hygiene, temperature, and moisture) that promote the growth of fungi in feed mills, equipment, and on farms (such as silos, automatic transport systems, and feeders). In addition to these preventive initiatives, mechanisms for analysis and monitoring must be established, particularly in feed mills that are contemporary and employ the most recent analytical methods (Berthiller et al. 2007). Small compounds (glycosides, glucuronides, fatty acid esters, and proteins) linked to mycotoxins can potentially obscure them from analytical detection, producing a false-negative result. As a result, traditional analytical techniques are unable to identify these hidden mycotoxins (Berthiller et al. 2005). In addition to preventive measures, analytical methods, and good hygiene, additional technologies could be useful in managing mycotoxicosis. These consist of the following: (1) different types of toxin binders; Ramos et al. (1996); (2) acidifiers (various acid combinations for use in feedstuffs and/or complete feed; Binder (2007); and (3) dis-activators of the various mycotoxins present in feed and/or feedstuffs (Volke et al. 2004).

8.11 Conclusion

Mycotoxicosis symptoms can manifest even at toxin concentrations below the limit of detection. The rapid expansion of knowledge on the impacts of mycotoxins is mostly due to the development of innovative analytical techniques. Negative analytical results do not rule out the possibility of mycotoxins in feedstuffs. Furthermore, mycotoxins are frequently not evenly distributed throughout the meal. Because of this, even with ideal sampling techniques, mycotoxins may remain analytically undetectable (Binder 2007). In many circumstances, clinical symptoms are not very noticeable. The farm's practitioner plays a crucial role in bringing attention to a potential mycotoxicosis issue that may be causing breeding pigs to experience reproductive failure. Problems can only be handled by implementing coordinated preventative measures and actions by all parties involved in pig health monitoring.

References

Alassane-Kpembi I, Schatzmayr G, Taranu I, Marin D, Puel O, Oswald IP (2016) Mycotoxins co-contamination: methodological aspects and biological relevance of combined toxicity studies. Crit Rev Food Sci Nutr. https://doi.org/10.1080/10408398.2016.1140632

Alexopoulos C (2001) Association of Fusarium myco-toxicosis with failure in applying an induction of parturition program with PGF2alpha and oxytocin in sows. Theriogenology 55:1745–1757

Alm H, Brussow K-P, Torner H, Vanselow J, Tomek W, Danicke S, Tiemann U (2006) Influence of Fusarium—toxin contaminated feed on initial quality and meiotic competence of gilts oocytes. Reprod Toxicol 22:44–50

Andretta I, Lovatto PA, Hauschild L, Dilkin P, Garcia GG, Lanferdini E, Cavazini NC, Mallmann CA (2008) Feeding of pre-pubertal gilts with diets containing zearalenone. Arq Bras Med Vet Zootec 60:1227–1233

Antonissen G, Martel A, Pasmans F, Ducatelle R, Verbrugghe E, Vandenbroucke V et al (2014) The impact of Fusarium mycotoxins on human and animal host susceptibility to infectious diseases. Toxins (Basel) 6:430e52

Barnikol H, Gruber S, Thalmann A, Schmidt HL (1982) Ergot poisoning in pigs (in German). Tieraertzliche Umschau 5:324–332

Berger T, Esbenshade KL, Diekman MA, Hoagland T, Tuite J (1981) Influence of prepubertal consumption of zearalenone on sexual development of boars. J Anim Sci 53:1559

Berthiller F, Dall'Astra C, Schuhmacher R, Lemmens M, Adam G, Krska R (2005) Masked mycotoxins: determination of deoxynivalenol glucoside in artificially and naturally contaminated wheat by liquid chromotagraphy-tandem mass spectrometry. J Agric Food Chem 53:3421–3425

Berthiller F, Sulyok M, Krska R, Schuhmacher R (2007) Chromatographic methods for the simultaneous determination of mycotoxins and their conjugates in cereals. Int J Food Microbiol 119:33–37

Binder EM (2007) Managing the risk of mycotoxins in modern feed production. Anim Feed Sci Technol 133:149–166

Booth NH, McDonald LE (1982) Veterinary pharmacology and therapeutics, 6th edn. Iowa State University Press, Iowa, pp 1093–1096

Burel C, Tanguy M, Guerre P, Boilletot E, Cariolet R, Queguiner M et al (2013) Effect of low dose of fumonisins on pig health: immune status, intestinal microbiota and sensitivity to Salmonella. Toxins (Basel) 5:841e64

Christensen CM, Mirocha CJ, Nelson GH, Quast JF (1972) Effect of young swine of consumption of rations containing corn invaded by Fusarium roseum. Appl Microbiol 23:202

Cysewski SJ, Wood RL, Pier AC, Baetz AL (1978) Effects of aflatoxin on the development of acquired immunity to swine erysipelas. Am J Vet Res 39:445e8

Danicke S, Doll S, Goyarts T, Valenta H, Ueberschar KH, Flachowsky G (2008) On the evaluation of the occurrence of the Fusarium-toxins deoxynivalenol (DON), zearalenone (ZON) and their metabolites in physiological substrates of the pig. Tieraerztl Prax Ausg Grosstiere Nutztiere 36:35–47

Desjardins AE (2006) Fusarium mycotoxins: chemistry, genetics and biology. APS Press, St. Paul, Minnesota, p 260

Diekman MA, Green ML, Malayer JR, Brandt KE, Long GG (1986) Effect of zearalenone and estradiol benzoate on serum LH and FSH in ovariectomized gilts. J Anim Sci 63(Suppl. 1):330

Diekman MA, Green ML (1992) Mycotoxins and reproduction in domestic livestock. J Anim Sci 70:1615–1627

Edwards S, Cantley TC, Rottinghaus GE, Osweiler GD, Day BN (1987a) The effects of zearalenone on reproduction in swine. I. The relationship between ingested zearalenone dose and anestrus in non-pregnant sexually mature gilts. Theriogenology 28:43–49

Edwards S, Cantley TC, Day BN (1987b) The effects of zearalenone on reproduction in swine. II. The effect on puberty attainment and postweaning rebreeding performance. Theriogenology 28:51–58

Etienne M, Jemmali M (1982) Effects of zearalenone (F2) on estrus activity and reproduction in gilts. J Anim Sci 55:1–10

Fink-Gremmels J, Malekinejad H (2007) Clinical effects and biochemical mechanisms associated with exposure to the mycoestrogen zearalenone. Anim Feed Sci Technol 137:326e41

Flowers B, Cantley T, Day BN (1987) A comparison of effects of zearalenone and estradiol benzoate on reproductive function during the estrus cycle in gilts. J Anim Sci 65:1576–1584

Furtado RM, Pearson AM, Hogberg MG, Miller ER, Gray JI, Aust SD (1982) Withdrawal time required for clearance of aflatoxins from pig tissues. Journal of Agriculture and Food Chemistry 30:101–106

Gan F, Zhang Z, Hu Z, Hesketh J, Xue H, Chen X et al (2015) Ochratoxin A promotes porcine circovirus type 2 replication in vitro and in vivo. Free Radic Biol Med 80:33e47

Glavits R, Sandor GS, Vanyi A, Gajdacs G (1983) Reproductive disorders caused by trichothecene mycotoxins in a large-scale pig herd. Acta Vet Hung 31:173–180

Grenier B, Oswald IP (2011) Mycotoxin co-contamination of foods and feeds: metaanalysis of publications describing toxicological interactions. World Mycotoxin J 4:285e313

Grenier B, Loureiro-Bracarense AP, Lucioli J, Pacheco GD, Cossalter AM, Moll WD et al (2011) Individual and combined effects of subclinical doses of deoxynivalenol and fumonisins in piglets. Mol Nutr Food Res 55:761e71

Hale OM, Wilson DM (1979) Performance of pigs on diets containing heated or unheated corn with or without aflatoxin. J Anim Sci 48:1394–1400

Hendel EG, Gott PN, Murugesan GR, Jenkins T (2017) Survey of mycotoxins in 2016 United States corn. J Anim Sci 95(Suppl. 4):16–17. https://doi.org/10.2527/asasann.2017.033

Jacela JY, DeRouchey JM, Tokach MD, Goodband RD, Nelssen JL, Renter DG, Dritz SS (2010) Feed additives for swine: fact sheets—flavors and mold inhibitors, mycotoxin binders, and antioxidants. J Swine Health Prod 18:27–32

Jacobson WC, Harmeyer WC, Jackson JE, Armbrecht B, Wiseman HG (1978) Transmission of aflatoxins B1 into the tissues of growing pigs. Bull Environ Contam Toxicol 19:156–161

Joens LA, Pier AC, Cutlip RC (1981) Effects of aflatoxin consumption on the clinical course of swine dysentery. Am J Vet Res 42:1170e2

Kopinski JS, Blaney BJ, Murray S-A, Downing JA (2007) Effect of feeding sorghum ergot to sows during mid lactation on plasma prolactin and litter performance. J Anim Physiol Anim Nutr 92:554–561

Kordic B, Pribicevic S, Muntanola-Cvetkovic M, Mikolic P, Nikolic B (1992) Experimental study of the effects of known quantities of zearalenone on swine reproduction. J Environ Pathol Toxicol Oncol 11:211–215

Koshinsky HA, Khachatourians GG (1992) Trichothecene synergism, additivity and antagonism: the significance of the maximally quiescent ratio. Nat Toxins 1:38–47

Lawlor PG, Lynch PB (2001) Mycotoxins in pig feeds—1: source of toxins, prevention and management of mycotoxicosis. Ir Vet J 54:117–120

Long GG, Diekman MA (1984) Effect of purified zearalenone on early gestation in gilts. J Anim Sci 59:1662–1670

Long GG, Diekman MA (1986) Characterization of effects of zearalenone in swine during early pregnancy. Am J Vet Res 47:184–187

Long GG, Diekman M, Tuite JF, Shannon GM, Vesonder RF (1982) Effect of *Fusarium roseum* corn culture containing zearalenone on early pregnancy in swine. Am J Vet Res 43:1599–1603

Long GG, Turek J, Diekman MA, Scheidt AB (1992) Effect of zearalenone on days 7 to 10 postmating on blastocyst development and endometrial morphology in sows. Vet Pathol 29:60–67

Meissonnier GM, Pinton P, Laffitte J, Cossalter AM, Gong YY, Wild CP et al (2008) Immunotoxicity of aflatoxin B1: impairment of the cell-mediated response to vaccine antigen and modulation of cytokine expression. Toxicol Appl Pharmacol 231:142e9

Meyer K, Usleber E, Martlbauer E, Bauer L (2000) Occurrence of zearalenone, alpha and beta—zearalenone in bile of breeding sows in relation to reproductive performances. Berliner und Munchener Tierartzliche Wochenschrift 113:374–379

Obremski K, Gajecki M, Zwierzchowski W, Zielonka L, Otrocka-Domagala I, Rotkiewicz T, Mikolajczyk A, Gajecki M, Polak M (2003) Influence of zearalenone on the reproductive system cell proliferation in gilts. Pol J Vet Sci 6:239–245

Osweiler GD, Ensley SM (2012) Mycotoxins in grains and feeds. In: Zimmerman JJ, Karriker LA, Ramirez A, Schwartz KJ, Stevenson GW (eds) Diseases of swine, 10th edn. John Wiley & Sons, Inc., Oxford, pp 938–952

Palyusik M, Harrach B, Mirocha CJ, Pathre SV (1980) Transmission of zearalenone and zearalenol into por cine milk. Acta Vet Acad Sci Hung 28:217–222

Pestka JJ, Tai JH, Witt MF, Dixon DE, Forsell JH (1987) Suppression of immune response in the B6C3F1 mouse after dietary exposure to the Fusarium mycotoxins deoxynivalenol (vomitoxin) and zearalenone. Food Chem Toxicol 25:297e304

Pestka JJ, Zhou HR, Moon Y, Chung YJ (2004) Cellular and molecular mechanisms for immune modulation by deoxynivalenol and other trichothecenes: unraveling a paradox. Toxicol Lett 153:61e73

Posa R, Magyar T, Stoev SD, Glavits R, Donko T, Repa I et al (2013) Use of computed tomography and histopathologic review for lung lesions produced by the interaction between Mycoplasma hyopneumoniae and fumonisin mycotoxins in pigs. Vet Pathol 50:971e9

Rafai P, Tuboly S, Bata A, Tilly P, Vanyi A, Papp Z, Jakab L, Tury E (1995) Effect of various levels of T-2 in the immune system of growing pigs. Vet Rec 136:511–514

Rainy MR, Tubbs RC, Bennett LW, Cox NM (1990) Prepubertal exposure to dietary zearalenone alters hypothalamohypophysial function but does not impair postpubertal reproductive function of gilts. J Anim Sci 68:2015–2022

Ramos AJ, Fink-Gremmels J, Hernandez E (1996) Prevention of toxic effects of mycotoxins by means of non nutritive absorbent compounds. J Food Prot 59:631–641

Ranzenigo G, Caloni F, Crernonesi F, Aad PY, Spicer LJ (2008) Effects of Fusarium mycotoxins on steroid production by porcine granulosa cells. Anim Reprod Sci 107:115–130

Rodrigues I, Naehrer K (2012) A three-year survey on the worldwide occurrence of mycotoxins in feedstuffs and feed. Toxins (Basel) 4:663e75

Ruhr LP, Osweiler GD, Foley CW (1983) Effect of the estrogenic mycotoxin zearalenone on reproductive potential in the boar. Am J Vet Res 44:483–485

Osweiler GD, Stahr HM, Beran GW (1990) Relationship of mycotoxins to swine reproductive failure. J Vet Diagn Invest 2:73–75

Savard C, Gagnon CA, Chorfi Y (2015a) Deoxynivalenol (DON) naturally contaminated feed impairs the immune response induced by porcine reproductive and respiratory syndrome virus (PRRSV) live attenuated vaccine. Vaccine 33:3881e6

Savard C, Provost C, Alvarez F, Pinilla V, Music N, Jacques M et al (2015b) Effect of deoxynivalenol (DON) mycotoxin on in vivo and in vitro porcine circovirus type 2 infections. Vet Microbiol 176:257e67

Silvotti L, Petterino C, Bonomi A, Cabassi E (1997) Immunotoxicological effects on piglets of feeding sows diets containing aflatoxins. Vet Rec 141:469–472

Smith TK, McMillan EG, Castillo JB (1997) Effect of feeding blends of fusarium mycotoxin-contaminated grains containing deoxynivalenol and fusaric acid on growth and feed consumption of immature swine. J Anim Sci 75:2184–2191

Smith TK, Az-Llano G (2009) A review of the effect of feed-borne mycotoxins on pig health and reproduction. Sustainable Animal Production—the Challenges and Potential Developments for Professional Farming:261–272

Stoev SD, Goundasheva D, Mirtcheva T, Mantle PG (2000) Susceptibility to secondary bacterial infections in growing pigs as an early response in ochratoxicosis. Exp Toxicol Pathol 52:287e96

Stoev SD, Gundasheva D, Zarkov I, Mircheva T, Zapryanova D, Denev S et al (2012) Experimental mycotoxic nephropathy in pigs provoked by a mouldy diet containing ochratoxin A and fumonisin B1. Exp Toxicol Pathol 64:733e41

Stubblefield RD, Honstead JP, Shotwell OL (1991) An analytical survey of aflatoxins in tissues from swine grown in regions reporting 1988 aflatoxin-contaminated corn. J Assoc Off Anal Chem 74:897–899

Streit E, Schatzmayr G, Tassis P, Tzika E, Marin D, Taranu I, et al. Current situation of mycotoxin contamination and co-occurrence in animal feed—focus on Europe. Toxins (Basel) 2012;4:788e809

Taranu I, Marin DE, Bouhet S, Pascale F, Bailly JD, Miller JD et al (2005) Mycotoxin fumonisin B1 alters the cytokine profile and decreases the vaccinal antibody titer in pigs. Toxicol Sci 84:301e7

Thieu NQ, Ogle B, Pettersson H (2008) Screening of aflatoxins and zearalenone in feedstuffs and complete feeds for pigs in Southern Vietnam. Trop Anim Health Prod 40:77–83

Tiemann U, Danicke S (2007) In vivo and in vitro effects of the mycotoxins zearalenone and deoxynivalenol on different non-reproductive and reproductive organs in female pigs: a review. Food Addit Contam 24:306–314

Tsakmakidis IA, Lymberopoulos AG, Vainas E, Boscos CM, Kyriakis SC, Alexopoulos C (2007) Study on the in vitro effect of zearalenone and alpha-zearalenol on boar sperm-zona pellucida interaction by hemizona assay application. J Appl Toxicol 27:498–505

Vandenbroucke V, Croubels S, Martel A, Verbrugghe E, Goossens J, Van Deun K et al (2011) The mycotoxin deoxynivalenol potentiates intestinal inflammation by Salmonella typhimurium in porcine ileal loops. PLoS One 6:e23871

Vanyi A, Glavits R, Gajdacs E, Sandor G, Jovacs F (1991) Changes induced in newborn piglets by the trichothecene toxin T-2. Acta Vet Hung 39:29–37

Venturini MC, Quiroga MA, Risso MA, Lorenzo CD, Omata Y, Venturini L et al (1996) Mycotoxin T-2 and aflatoxin B1 as immunosuppressors in mice chronically infected with toxoplasma gondii. J Comp Pathol 115:229e37

Volke A, Vogler B, Schollenberger M, Karlovsky P (2004) Microbiological detoxification of mycotoxin desoxynivalenol. J Basic Microbiol 44:147–156

Weaver GA, Kurtz HJ, Mirocha CJ, Bates FY, Behrens JC, Robinson TS, Gipp WF (1978a) Mycotoxin-induced abortions in swine. Can Vet J 19:72–74

Weaver GA, Kurtz HJ, Mirocha CJ, Bates FY, Behrens JC, Robinson TS (1978b) Effect of T-2 toxin on porcine reproduction. Can Vet J 19:310–314

Weaver GA, Kurtz HJ, Bates FY, Chi MS, Mirocha CJ, Behrens JC, Robison TS (1978c) Acute and chronic toxicity of T-2 mycotoxin in swine. Vet Rec 103:531–535

Young LG, King GJ (1983) Prolonged feeding of low levels of zearalenone to young boars. J Anim Sci 57(Suppl. 1):313–314

Young LG, King GJ (1986) Low concentrations of zearalenone in diets of mature gilts. J Anim Sci 63:1191–1196

Young LG, Vesonder RF, Funnell HS, Simons I, Wilcock B (1981) Moldy corn in diets of swine. J Anim Sci 52:1312–1318

Young LG, Ping H, King GJ (1990) Effects of feeding zearalenone to sows on rebreeding and pregnancy. J Anim Sci 68:15–20

Effect of Environment on Pig's Health

Amitava Roy and Tanmoy Rana

Abstract

Ensuring animal well-being on farms necessitates good husbandry practices. The Directive 2008/120/EC, one of the most significant legal instruments governing pig husbandry practices, stipulates that group-housed pigs must have access to litter or other items that allow for occupation and exploration. The Commission Recommendation (EU) 2016/336 on the application of Council Directive 2008/120/EC, which was published in 2016, describes the different types of materials that can be utilised to enhance the welfare of animals. Straw is regarded as the best material for pig housing, according to the paper; however, materials that are classified as suboptimal—like wood bark—and materials of minor interest—like plastic toys—are frequently utilised in scientific study and practical applications. Therefore, the purpose of this work is to examine and organise the existing body of evidence about the effects of environmental enrichment on the welfare of pigs. The main concerns raised in this chapter are the effects of different enrichment methods on the decrease of aggressive behaviour, tail biting, and stereotyping during the pre-, post-, and fattening stages of pig production.

Keywords

Enrichment · Pigs · Welfare

A. Roy (✉)
Department of Livestock Farm Complex, West Bengal University of Animal & Fishery Sciences, Kolkata, India

T. Rana
Department of Veterinary Clinical Complex, West Bengal University of Animal and Fishery Sciences, Kolkata, West Bengal, India

© The Author(s), under exclusive license to Springer Nature Singapore Pte Ltd. 2024
T. Rana, B. Soto-Blanco (eds.), *Good Practices and Principles in Pig Farming*, Livestock Diseases and Management,
https://doi.org/10.1007/978-981-97-4665-1_9

9.1 Introduction

The food production industry has seen an increase in customer interest in recent years. When considering products derived from animals, it can be said that the public discourse is not just concerned with the items' high quality. Currently, keeping farms in proper husbandry conditions has also received a lot of attention. Qualities of animal welfare have long been recognised (Fraser et al. 1997). This phrase has been associated with numerous matters mostly concerning the provision of a suitable physical and mental state for animals. Performing invasive procedures on farm animals, like tail docking, teeth trimming, and castration, is highly controversial these days—especially when done without anaesthesia (Godyń et al. 2013). Legislative changes frequently reflect the pressure exerted on these and other areas of animal rights by animal welfare organisations and consumers. However, even experts, government inspectors, advisors, and knowledgeable people frequently lack the knowledge necessary to pinpoint the reasons behind tail biting, suggest ways to stop it, and comprehend relevant laws (Hothersal et al. 2016). The Directive 2008/120/EC (EC Directive 2008/120/EC 2019) is one of the most significant legal regulations that establishes the minimum criteria for pigs. Among other things, this law specifies that pigs housed in groups must have access to litter or other materials that allow them to explore and make themselves comfortable. Aggression and other abnormal behaviour can be decreased by improving the pigs' surroundings (EFSA 2012). Tail biting can be influenced by a number of factors, including an unsuitable diet, delayed or absent food provision, gastrointestinal distress, poor health, genotype, excessive density, and unfavourable microclimate circumstances. The absence of a substrate or manipulable object in the pigs' environment is another significant aspect that can contribute to a higher frequency of this behaviour (Taylor et al. 2010). Regarding the steps required to lessen the need for tail-docking, the Commission Recommendation (EU) 2016/336 on the application of Council Directive 2008/120/EC laying out the basic criteria for the protection of pigs was released in 2016. As mentioned in the document, this process should never be carried out on a regular basis. This chapter explains and characterises the numerous categories of materials that may be used to improve animal well-being, more so than the regulation (EC regulation 2008/120/EC, 2019), taking into account the element of enhancing the pigs' surroundings.

According to the Commission Recommendation's four-point rating system, enrichment products should be manipulable, chewable, palatable, and investigable. Furthermore, the text stipulates that enrichment materials must be supplied in a manner that makes them sustainable (i.e. consistently restocked), accessible for oral manipulation, and sufficiently abundant for pigs. They ought to be tidy as well. The materials can only be deemed ideal if they satisfy every one of the aforementioned requirements. Since subpar materials have most of the aforementioned qualities, they ought to be used in pig housing in conjunction with other materials. Materials of minor interest make up the final category of enrichments. Since they are unable to meet every need of the animal, they must be used in conjunction with both optimal and inferior materials. Straw is difficult to introduce into slatted floor systems because it can clog the manure system, even though it is thought to be the best material for improving pig housing (based on the qualities of various enrichment materials included in the recommendation) (Wallgren et al. 2016). However, enrichment devices composed of plastic and metal that are frequently employed in conventional farming may only be classified as materials of marginal relevance. As a result, developing and implementing novel materials and techniques for environmental enrichment in pig housing is imperative. Analysing study findings about the effects of different enrichment devices and confirming that the materials used in the trials adhere to the Commission Recommendation's specifications seem to be equally significant. This chapter's objective is to examine and organise the available data on the effect of environmental enrichment on pig well-being. This chapter addresses the challenges of fattening pigs, pre- and post-weaning piglets, and housing diverse pig groups more effectively using the available materials. The most recent research has been incorporated into this chapter, mostly focusing on how enrichment materials might lessen negative behaviours like aggression, tail biting, and stereotyping.

9.2 Pig Welfare Issues Associated with Intensive Production Systems and the Welfare Assessment Techniques

9.2.1 The Idea of Animal Welfare and Evaluation Techniques

For a long time, animal well-being has been a scientific idea (Fraser 2008). It can range from really low to very good can be assessed using scientific methods (Broom 2011). The concept of animal welfare encompasses various aspects, including the significance of the animal's capacity to maintain equilibrium in its body and mind under varying environmental circumstances (Broom 1988). According to Broom (1986), welfare is the state in which a person is making an effort to adapt to their surroundings. When coping mechanisms involving behavioural, physiological, immunological, and other brain-coordinated components don't work, animals suffer from poor welfare. In recent years, it has been normal practice to employ animal well-being indicators that are based on behavioural changes and physiology (Broom 2011). Animal well-being is impacted by a variety of factors, including the existence of sickness, injury, social interaction within the herd, housing conditions, and human handling (Broom 2011). Maintaining the animal's mental and physical stability may be difficult under inappropriate social, sanitary, and microclimatic settings. Even when a person can adjust to difficult circumstances, they can nevertheless suffer—experience discomfort or annoyance. Therefore, excellent welfare is not always synonymous with adaptation (Boissy et al. 2007). Positive welfare indicators are becoming more and more common at the moment (McCormick 2012). McCormick (McCormick 2012) presented three behavioural indicators as a means of assessing welfare: the indicators of luxury habits, the indicators of contentment/pleasure, and the indications of behaviours that enhance the capacity to handle adversity. Additionally, play, barking, and tail movement assessment can be used to identify happy emotions in pigs, whereas ear movements, low tail, freezing, urinating, defecating, trying to flee, and high-pitched vocalisations can suggest negative emotions (Reimert et al. 2013).

Unwanted behaviour can be viewed as a sign of poor welfare when it occurs (Mason and Rushen 2006). The scoring of wounds and skin damage is a crucial tool in evaluating animal welfare (Turner et al. 2006). Pigs' front body lesions are a reflection of how frequently they fight with each other in the pen. Conversely, there is no evidence linking the development of this behaviour to skin injury in the central area of the pig's body. In addition to behavioural indications, physiological parameters are also employed in the evaluation of animal well-being. Stress hormone assessments are a common method of measuring physiological indices of animal welfare in animal studies (Mormède et al. 2007). An animal's body responds to inappropriate environmental stimuli by altering the hypothalamo-pituitary-adrenocortical (HPA) axis, which results in the release of glucocorticoids among other things (Morris et al. 2012). One may argue that chronic stress, as opposed to acute stress, is a more pressing problem when it comes to animal welfare. An animal experiences chronic stress when it is unable to adapt to its surroundings. It is noteworthy to mention that Munsterhjelm et al. (Munsterhjelm et al. 2010) found that chronic stress symptoms are linked to a richer environment as opposed to a barren one. According to Keay et al. (2006), a prolonged increase in cortisol levels causes impairment and degradation of the emotional and physical states. Although measuring cortisol in serum, or blood, has been a common method for evaluating animal health, there is now a search on for less invasive techniques than blood sample (Kersey and Dehnhard 2014). Stress hormones can be secreted by animals as a result of handling and immobilisation, which can be perceived as a danger and a cause of worry (Stewart et al. 2005). The hypothalamic-pituitary-adrenal (HPA) axis can also be reliably activated by measuring glucocorticoids in saliva, faeces, or hair samples, and the process of collecting these materials is less invasive (Casal et al. 2017). But since the body's various systems are involved in stress responses, measuring cortisol levels is only one aspect of the physiological characteristics that need to be assessed (Blache et al. 2007). The Mkwanazi et al. (2019) assessment brings up a number of concerns about different types of enrichments in pig housing. The aforementioned authors covered, among other things, blood metabolites and performance in pigs housed in enriched and barren pens. According to certain data, environmental enrichment may increase growth rate and decrease creatine kinase (Peeters and Geers 2006). The use of a variety of behavioural techniques in conjunction with non-invasive biomarkers is crucial for the accurate evaluation of animal welfare.

9.2.2 The Needs and Natural Behaviour of Pigs

It is important to note that animals have powerful motivating mechanisms to ensure that their basic requirements are met while talking about ideas related to animal well-being. In addition to their fundamental needs for food and drink, these also include the freedom to explore their environment and exhibit their natural

Fig. 9.1 Pigs' needs their innate behaviour

behaviours (Kittawornat and Zimmerman 2010). One of the most crucial tasks for pigs is to root; piglets root to obtain food, sows root to construct a nest, and rooting may also serve some thermoregulatory purposes (Burne et al. 2001). The pigs' attention appears to be drawn to other animals in the pen when there are no materials available for rooting (Fig. 9.1), foraging, or manipulating (Kelly et al. 2000). Some writers claim that the majority of pigs raised in Europe are kept in harsh, slatted-floor environments under conditions of intensive production (Guy et al. 2013). The display of behaviour peculiar to a species is not encouraged by this kind of housing. Research has demonstrated that under intensive production, limiting opportunities for environmental exploration might result in a rise in instances of violence, cannibalism, tail-biting, and stereotyping (Scott et al. 2006).

Play is seen as a luxury behaviour and, as such, a sign of well-being (Reimert et al. 2013). It is crucial, particularly for piglets who are young. Giving the piglets toys to play within their surroundings could help them become more social. According to some research, animals raised in a setting that encourages play behaviour are more equipped to handle challenging circumstances later in life (Spînka et al. 2001).

9.2.3 Risk Elements for Adverse Behaviours in Severe Situations

A major issue for the welfare and financial sides of pig raising is tail biting. Three distinct types of tail biting were recognised by Taylor et al. (2010): two-stage, sudden-forceful, and compulsive. Pigs are known to participate in a "routine" called "two-stage tail biting," in which the victim receives skin damage, when they gently touch another person's tail. According to Schrøder-Petersen and Simonsen (2001), the presence of fresh blood may entice other pigs to bite, which could ultimately result in increased aggression and, in extreme circumstances, cannibalism within the herd. By offering some things to foster a manipulative interest, this type of tail biting may be avoided. An acute and quick attack carried out without any prior mild tail manipulation is one way to describe another variation of this behaviour. It frequently occurs when pigs must fight for food or water (Morrison et al. 2007). The majority of those who engage in compulsive tail biting are one or a small group of persons. Tail biters appear to be selective about this area of the body and are always on the lookout for another to bite or grip. Obsessive tail biting may be interpreted as a type of stereotyping, and the pathology's development may be associated with inappropriate nutrition and genetic lines.

When individuals from various herds are put together in a single pen for mixing, which is widely seen as a stressful treatment, an increase in aggressive behaviour is sometimes noted. Because it is necessary to establish the social hierarchy, violence and aggressiveness are frequently associated with it (McGlone 1986). Individual fights within the pen eventually decrease until after the establishment of the herd hierarchy (Parratt et al. 2006). The belly nosing phenomenon, which is defined as a piglet rhythmically rubbing its nose on another piglet's belly, is another type of behaviour that is primarily seen in weaned piglets, particularly in those who were weaned earlier (Fraser 1978). A detailed description of the numerous aspects associated with the occurrence of this stereotypical behaviour was provided by Widowski et al. (2008). Early separation from the sow is one of the causes of belly suckling or belly nosing. When piglets engage in this behaviour over an extended length of time, it frequently results in lesions and can have a detrimental effect on the pigs' performance (Widowski et al. 2008). These authors provided some data regarding the intricate relationship between weaned piglets' belly nosing, drinking, feeding, and nursing.

In short, early weaned piglets' intense desire to nurse combined with their underdeveloped capacity to feed themselves may be the cause of oral behavioural issues. Belly nosing may possibly have a hereditary foundation or be a general behavioural response to stressful situations, as Breuer et al. (2003) suggested (Salak-Johnson et al. 2004). Poor stimuli in the environment could potentially have a role in the development of this behaviour (Cooper et al. 2001). Research has shown that farm, lab, and wild animals housed in arid environments can develop stereotypical behaviour (Mallapur and Cheelam 2002). According to Cronin and Wiepkema (1984), stereotypes are characterised as basic ritualised behaviours that are repetitive and non-functional. It is frustrating when an animal is very motivated to satisfy its wants yet is unable to do so. Stereotypes or replacement behaviour may be manifestations of this frustration (Mason 1991). Additionally, some stereotypes may be an attempt to minimize the negative bodily effects of stress. As a result of the animal's attention being drawn away from the conflict's cause, this behaviour may result in decreased anxiety and an increased capacity to respond to damaging external stimuli (Mason 1991). Stereotypes are mostly associated with basal ganglia function. In summary, it may be presumed that behaviours linked to unceasing repetition stem from a variety of factors, including an imbalance in the activation of the two primary basal ganglia regulatory loops (Langen et al. 2011).

Fig. 9.2 Improving the environment as an effective prevention and mitigation strategy

9.2.4 Enhancing the Environment as a Successful Tail-Biting Prevention and Mitigation Method

Additionally, it has been demonstrated in animal models that the prevention of stereotypies through contextual enrichment is associated with increased levels of dendritic morphology (i.e. more dendritic spines) and neuronal metabolic activity in the brain's motor areas (Turner and Lewis 2003). Given the complex and multifaceted nature of tail-biting risk, providing high-quality enrichment is a crucial first step, but it might not be sufficient to stop an epidemic of tail-biting (Arey and Franklin 1995) (Fig. 9.2). Pigs' involvement in interacting with their surroundings should be taken into consideration when evaluating their degree of well-being. It's important to see if the pigs can exhibit acceptable manipulative and exploration behaviours, if the enrichment material is palatable, chewable, and breakable, and if the pigs interact more with playpen components than with enrichment objects or materials. The directive states that it's also critical to ascertain if every pig has ongoing access to an adequate supply of clean, safe enrichment materials. The best options for environmental enrichment may be chosen using the results of this evaluation as a guide.

9.3 Various Forms of Enhancement

In the 1940s, the concept of environmental enrichment for experimental animals was initially put forth (Hebb 1947). Since then, a great deal of research has been done on the beneficial effects of improving the animal's surroundings on its brain structure and biochemical alterations (Reynolds et al. 2010). The main principle behind environmental enrichment for animals is that the objects they interact with should be novel in some way, stimulating their visual, somatosensory, and olfactory systems (Nithianantharajah and Hannan 2006). Given that pigs might get disinterested in an object after just a few days (Ernst et al. 2018), it is crucial to maintain the animals' interest by regular replenishment or renewal of enrichments. In addition, there ought to be enough of toys for the pigs to play with simultaneously. The aforementioned Commission Recommendation (EU) 2016/336 (2016) contained these explanatory notes. Straw is one of the best materials for animal welfare when it comes to bedding, but other materials like green fodder, chopped or crushed miscanthus, and root vegetables may also be optimal depending on certain types of enrichment (The European Commission 2016). According to Van de Weerd et al. (2003), pigs are first drawn to objects with scent, chewability, and deformability; nevertheless, the enrichment's destructibility and edibility may draw them in for a longer period of time. Peanut shells, fresh wood, corn cobs, natural ropes, compressed straw cylinders, shredded paper, pellets, and other materials are deemed suboptimal in the Commission Recommendation. Furthermore, straw that is placed in dispensers or racks is regarded as a subpar material (EU 2016). Among the marginal materials are balls, pipes, hard woods, rubber, soft plastic, chains, and rubber. When bedding is not available, a combination of several enrichment materials should be utilised for optimal effects (EU 2016).

Even though the Council Directive 2008/120/EU was established many years ago, it appears that some tests are still being conducted utilising materials and items that are not allowed by this formal EU Act, and the results are indicating some good effects on the behaviour of pigs. However, prior research by Scott et al. (2007) and Zwicker et al. (2012) shows that, even with an increase in the number of hanging objects, straw bedding far more effectively ensures a high interest in this sort of enrichment than hanging toy(s). Furthermore, the provision of straw in racks encourages explorers' behaviour, particularly when additional such enrichment is offered. Because hanging elements are not rootable and most of them are not edible, they do not meet the recommendation's requirements and do not offer pigs as much interest as straw. It is noteworthy to mention that the Recommendation has a lot more important details than the Directive. A chapter covering most of the many solutions utilised for environmental enrichment in conventional pig raising is covered in the Ernst et al. (2018) publication. Several writers attest to the advantages of utilising several enrichments concurrently (Ernst et al. 2018). Prior to the implementation of Commission Recommendation 336/2016, the most often utilised objects were commercially available plastic or rubber toys or metal chains (Bracke et al. 2012). Some evidence suggests that this might still be the case. For instance, the final report of an audit conducted in Germany between 12 February 2018 and 21 February 2018 to assess member state initiatives to stop tail-biting and steer clear of routinely tail-docking pigs (European Commission 2018) revealed that the German authorities have the authority to levy administrative fines for directive non-compliance.

However, it was observed that, with regard to the qualities and adequate amounts of enrichment material, certain of the requirements of Council Directive 2008/120/EC are not immediately sanctionable through this method. Furthermore, despite significant financial investments made by regional and central authorities for research and results dissemination, their efforts to curtail tail biting and stop routine pig tail-docking have yielded little noticeable progress, and tail-docking continues to occur in Germany. Tail-docking is estimated by the central competent authority to occur more than 95% of the time in Germany. Furthermore, chains are regarded as "acceptable" as enrichment by the French Agency for Food, Environmental, and Occupational Health and Safety (Laloy and Huet 2017). Environmental enrichment can include bigger spaces, scents, or even music in addition to the best items and materials specified in the Commission Recommendation (Martin et al. 2015). Food can be a unique kind of enrichment, and the way it is given to animals—for example, by hiding it in the substrate—may have some important advantages for raising the activity levels of pigs (Reimert et al. 2013). These days, pig welfare may benefit greatly from a variety of cognitive enrichment programmes, most of which are associated with an animal's acquisition of a reward (Ernst et al. 2005). Reimert et al. (2013) discovered that pigs' increased play behaviour is frequently associated with rewarding events that occur during training and testing. The authors also noticed that pigs are more likely to bark and wag their tails when they receive rewards. As a result, it is commonly accepted that an animal's reward is connected to producing happy feelings. It is important to always keep in mind, nevertheless, that legal criteria must be met.

9.4 Enhancement of the Environment for Weaned and Sucker Piglets

Piglets that have been pre- and post-weaned have been used to study the effects of environmental improvements. It appears that early exposure to enrichment in neonatal living settings may result in improved social behaviour in the pigs later on (Day et al. 2002). Research indicates that providing piglets with access to wood shavings, bark, straw, or even bits of newspaper during the pre-weaning phase may positively impact their behaviour (Yang et al. 2018). Early straw availability may have a positive impact on reducing tail-biting and nosing (Day et al. 2002). Having access to straw during the pre-weaning phase may also help piglets grow more healthily and reduce mounting and oral abuse of other people (Van Dixhoorn et al. 2016). Furthermore, Brajon et al. (2017) found that piglets with access to straw during the pre-weaning phase spent more time lying down, which might be interpreted as a sign of the animals' contentment.

When materials other than straw are available, positive signs of animal wellbeing can also be shown. As previously noted, there are several factors that contribute to belly nosing, which is typically regarded as a sign of low welfare in weaned piglets (Brajon et al. 2017). It has been previously reported that providing straw might not be sufficient to completely eradicate this occurrence (Statham et al. 2011). However, using black foam rubber matting was found to have some benefits in terms of reducing belly nosing (Bench and Gonyou 2006). Comparably, materials categorised as poor or of minor interest based on recommendations had the greatest impact on some behavioural and physiological variables in earlier studies (Telkänranta et al. 2014). According to Yang et al. (2018), piglets in the two groups with access to enrichment—hanging objects or wood bark—engaged in play behaviour more frequently than those housed in farrowing pens with no improvement (Fig. 9.3).

Conversely, every experimental animal in the study by Telkänranta et al. (2014) was kept in pens that had enriched environments. But the piglets in the experimental group had access to sisal ropes, plastic balls, newspaper fragments, and wood shavings, while the individuals in the control group received simply balls and wood shavings. Particularly the newspaper pieces and sisal ropes stimulated the pigs' activity and reduced their oral-nasal manipulation of their pen pals. Several studies have demonstrated further benefits of environmental enrichment during the newborn period (Statham et al. 2011). Chronic stress is known to have an immunosuppressive effect (Connor et al. 2005). It is worthwhile to consult the research of Yang et al. (2018) and Van Dixhoorn et al. (2016) when taking this factor into account. The aforementioned authors discovered that following weaning, pigs with access to wood bark during their neonatal period had reduced cortisol levels. Conversely, the latter authors discovered that piglets housed in a pre-weaning pen containing straw, damp peat, and wood shavings had a less severe beginning of co-infection with Actinobacillus pleuropneumoniae and the reproductive and respiratory virus later in life.

Conversely, Telkänranta et al. (2014) found that less severe tail damage—that is, injuries including swelling and infection or partial or complete loss of the tail—was brought on by higher environmental enrichment during the neonatal period. The study conducted by Yang et al. (2018) did not see this impact. Although there was

Fig. 9.3 Improvement of the weaned piglets' environment

no discernible effect on aggressive behaviour, the degree of hostility was assessed using the skin lesion before weaning and 2 days after mixing. Maybe a few more days of observation were necessary, though. Piglets housed in better (bigger and furnished with straw) and standard neonatal housing settings were compared in terms of behaviour by Martin et al. (2015). According to these authors, the hostility exhibited 3 days after weaning was similar whether or not enrichment was given, but after 7 days, the enriched group showed a marked decrease in this behaviour.

9.5 Enhancement of the Environment for Pig Fattening

Given the environmental enrichment involved in developing and completing pigs, it appears that a greater room is essential for ensuring animal comfort. Even with access to enrichment (cylindrical pieces of hard wood strung on a chain), the pigs housed in larger enclosures in the study conducted by Cornale et al. (2015) had higher levels of corticosteroids in their faeces. Conversely, the Di Martino et al. (2015) experiment shown that even in circumstances when the animals had access to straw racks, there was a high incidence of tail lesions under high stocking density settings. These findings would imply that having more space is a crucial component in potentially lessening this disease. It is important to note that fattening hefty pigs used in some operations need more area than what is required by law. These circumstances can arise when pigs weigh between 180 kg and 190 kg and when the fattening process takes a long time (Di Martino et al. 2015). According to Council Directive 2008/120/EC, pigs should have access to an environment that meets their demands for exercise and exploratory behaviour. Space constraints appear to be a serious threat to pig well-being. For pigs weighing up than 150 kg, the allotted 1 m^2 area seemed to be inadequate. Furthermore, the regulation specifies that other steps, including consideration of the habitat and stocking densities, should be done to prevent vices like as tail-biting before performing tail docking. It is necessary to alter insufficient environmental conditions or management systems as a result.

Tail biting incidence is also significantly influenced by one's genetic background (EFSA 2007). The research of Bulens et al. (2016) supported this. The authors examined the effects of completing pigs with two different genetic backgrounds using two different enrichment strategies: straw block and chain and solely chain provision. Regardless of the enrichment offered, the kind of boar was the main element in tail biting frequency. The Telkänranta et al. (2014) study indicated that the prevalence of moderate tail damage and ear biting decrease had some favourable characteristics when pieces of newly cut birch trees were provided. Although the authors also evaluated other materials, this enrichment proved to be the most beneficial in the previously described features. The control group's pens were furnished with a straw dispenser, wood shavings, and metal chain; the other four groups received the same enrichments along with additional items like a piece of fresh birch wood suspended horizontally, polythene pipes, and vertically suspended metal chains.

The final group was kept in pens furnished with every enrichment item previously listed. The study also revealed that, among other enrichment devices, freshly cut birch wood and polythene pipe had a substantial impact on the frequency of object manipulation. Straw continues to pique interest in research; a study by Scott et al. (2007) using this substrate as bedding showed this. Nannoni et al. (2018) found that finishing pigs in all production cycles showed interest in investigating the environmental enrichment within the racks, even in the absence of bedding (Fig. 9.4). In place of hanging chains, these writers examined hardwood logs and an edible vegetable block positioned inside metal racks. The statistical analysis revealed significant differences only in the moderate tail lesion observed in the group with access to the wooden logs compared to the group with the chain, taking into account the effect of enrichment materials on tail biting cases. It appears that the vegetal edible block was unable to completely stop the pigs' attention from being focused on other pigs, even though the authors claimed it helped to lessen the exploratory behaviour that was directed towards the corral floor. Some moderate to severe tail lesions were present in the group that had access to this enrichment. Competition for enrichment access may be the cause of this, as demonstrated by Bulens et al.'s (2016) study. In this experiment, the pigs with access to the straw dispenser had more mounting and fighting at the start of the finishing phase. Research by Casal et al. (2017) and Nannoni et al. (2018) examined the impact of enrichment on stress response in finishing pigs. The experimental groups in the Casal et al. (2017) study were housed in pens with sawdust, rubber balls, and natural hemp ropes, whereas the control group was kept in a barren setting. Supplementing with herbs was also tested by the writers. Following 2 months of therapy, the control group had the highest salivary chromogranin A (CgA) and hair cortisol levels. As a result, the authors propose that pigs may experience less stress as a result of both enrichment and

Fig. 9.4 Improvement of the pig fattening environment

herbal substances. Nevertheless, Nannoni et al. (2018) study's enrichment tools had no effect on the cortisol levels in the hair.

A lengthy study produced the results (Nannoni et al. 2018) that were previously mentioned. Regarding the amount of time spent exploring the surroundings, nuzzling, and associating with other pigs, there were no changes between the treatments. Additionally, agnostic behaviour, which was typically rather low during the trial period, was unaffected by the items. The increase in animal activity, even when only the content categorised by the EU Recommendation as being of minor interest, is delivered, was one of the primary findings of this study. The animals in the control group, who were denied access to a "toy," stayed motionless for the most part of the session. Based on these and other findings in this section, it can be concluded that, from the standpoint of animal welfare, an increase in animal activity and interest in an enrichment object is highly valuable, even though harmful behaviour persisted in the pigs housed with access to enrichment. Furthermore, Scott et al. (2007) noted that utilising straw was more environmentally beneficial than hanging toys; nevertheless, Telkänranta et al. (2014) found that new wood attracted pigs the most, even when they also had access to straw. These kinds of findings hold promise for the housing of pigs, when the limited or non-existent availability of straw is caused by the technological maintenance system. However, it should be noted that enrichment materials designated as being of marginal relevance cannot, in fact, be used alone in EU member states (without supply of optimal enrichment).

9.6 Conclusion

It may be concluded from the several research findings included in the chapter that there are numerous advantages to animal welfare from the enrichments given to pigs who are growing and young. In general, it may be said that focused activity is increased when there is even a tiny amount of substrate and objects available. Enrichment provided at the neonatal stage (farrowing pen) has been shown to enhance manipulating abilities and may also result in improved social behaviour displayed later in the animal's life. The primary outcome of enrichment is a decrease in the amount of focus on other pen members. This includes the pigs' propensity for biting, belly nosing, and tail gnawing. Nevertheless, there is conflicting evidence regarding the outcomes of reducing these negative behaviours by offering enrichments. Studies have supported the benefits of utilising multiple enrichments concurrently, but other findings have indicated that the intended behavioural change in the pigs could only be achieved by utilising materials that, in accordance with Commission Recommendation (EU) 2016/336, are deemed to be of marginal interest. The recommendation that "materials of marginal interest—materials providing a distraction for pigs, which should not be considered as fulfilling their essential needs, and therefore optimal or suboptimal materials should also be provided" can be deceptive for farmers because it emphasises this point strongly. Furthermore, a few writers discovered that materials with protruding tubes, like newsprint or PCV pipes, had a beneficial effect. Therefore, it is still unclear if these enrichments—as well as hard wood, whose gnawing may also cause harm—can be deemed entirely safe for the health of animals.

The European legislation governing the kinds of environmental enrichment materials and their characteristics cannot be changed, nor would it make sense. The present EU regulations reflect the outcomes of scientific studies conducted by the European Union and are based on research findings. It should always be remembered that ear and tail biting have complex origins that are frequently challenging to identify. Tail-docking should only be used as a last resort because it is illegal in some EU nations. Therefore, it is important to explore scientific and practical alternatives to halt and prevent tail biting, always adhering to legal laws. Tail-docking should be done as little as possible or not at all if the pigs' environment is altered, such as by lowering the density of their stockings and introducing the best environmental enrichment options based on scientific assessments of their efficacy.

References

Arey DS, Franklin MF (1995) Effects of straw and unfamiliarity on fighting between newly mixed growing pigs. Appl Anim Behav Sci 45:23–30
Bench CJ, Gonyou HW (2006) Effect of environmental enrichment at two stages of development on belly nosing in piglets weaned at fourteen days. J Anim Sci 84:3397–3403
Blache D, Terlouw C, Maloney SK (2007) Physiology. In: Appleby MC, Mench JA, Olsson IAS, Hughes BO (eds) Animal welfare, 2nd edn. CABI, Cambridge, pp 155–182
Boissy A, Manteuffel G, Jensen MB, Moe RO, Spruijt B, Keeling LJ, Winckler C, Forkman B, Dimitrov I, Langbein J et al (2007) Assessment of positive emotions in animals to improve their welfare. Physiol Behav 92:375–397
Bracke MBM, de Lauwere CCDC, Wind SM, Zonderland JJ (2012) Attitudes of Dutch pig farmers towards tail biting and tail docking. J Agric Environ Ethics 26:847–868
Brajon S, Ringgenberg N, Torrey S, Bergeron R, Devillers N (2017) Impact of prenatal stress and environmental enrichment prior to weaning on activity and social behaviour of piglets (*Sus scrofa*). Appl Anim Behav Sci 197:15–23
Breuer K, Sutcliffe MEM, Mercer J, Rance K, Beattie V, Sneddon I, Edwards S (2003) The effect of breed on the development of adverse social behaviours in pigs. Appl Anim Behav Sci 84:59–74
Broom DM (1986) Indicators of poor welfare. Br Vet J 142:524–526
Broom DM (1988) The scientific assessment of animal welfare. Appl Anim Behav Sci 20:5–19
Broom DM (2011) A history of animal welfare science. Acta Biotheor 59:121–137
Bulens A, Van Beirendonck S, Thielen J, Buys N, Driessen B (2016) Long-term effects of straw blocks in pens with finishing pigs and the interaction with boar type. Appl Anim Behav Sci 176:6–11
Burne THJ, Murfitt PJE, Gilbert CL (2001) Influence of environmental temperature on PGF2α-induced nest building in female pigs. Appl Anim Behav Sci 71:293–304
Casal N, Manteca X, Escribano D, Cerón JJ, Fàbrega E (2017) Effect of environmental enrichment and herbal compound supplementation on physiological stress indicators (chromogranin A, cortisol and tumour necrosis factor-α) in growing pigs. Animal 11:1228–1236
Connor TJ, Brewer C, Kelly JP, Harkin A (2005) Acute stress suppresses pro-inflammatory cytokines TNF-α and IL-1β independent of a catecholamine driven increase in IL-10 production. J Neuroimmunol 159:119–128
Cooper JJ, Cox LN, Whitworth C (2001) Early environmental experience and transferable skills in the weaned piglet. Anim Welf Potters Bar 10:S238

Cornale P, Macchi E, Miretti S, Renna M, Lussiana C, Perona G, Mimosi A (2015) Effects of stocking density and environmental enrichment on behavior and fecal corticosteroid levels of pigs under commercial farm conditions. J Vet Behav 10:569–571

Cronin GM, Wiepkema PR (1984) An analysis of stereotyped behaviour in tethered sows. Ann Rech Vet 15:263–270

Day JEL, Burfoot A, Docking CM, Whittaker X, Spoolder HAM, Edwards SA (2002) The effects of prior experience of straw and the level of straw provision on the behaviour of growing pigs. Appl Anim Behav Sci 76:189–202

Di Martino G, Scollo A, Gottardo F, Stefani AL, Schiavon E, Capello K, Marangon S, Bonfanti L (2015) The effect of tail docking on the welfare of pigs housed under challenging conditions. Livest Sci 173:78–86

EC Directive 2008/120/EC. Available online: https://eur-lex.europa.eu/legal-content/EN/TXT/PDF/?uri=CELEX:32008L0120&from=EN (accessed on 16 April 2019)

EFSA—Panel on Animal Health and Welfare (2007) The risks associated with tail biting in pigs and possible means to reduce the need for tail docking considering the different housing and husbandry systems. EFSA J 5:611

EFSA Panel on Animal Health and Welfare (2012) Statement on the use of animal-based measures to assess the welfare of animals. EFSA J 10:2767

Ernst K, Puppe B, Schön PC, Manteuffel G (2005) A complex automatic feeding system for pigs aimed to induce successful behavioural coping by cognitive adaptation. Appl Anim Behav Sci 91:205–218

Ernst K, Ekkelboom M, Kerssen N, Smeets S, Sun Y, Yin X (2018) Play behavior and environmental enrichment in pigs. WUR:1–59. Available online: https://www.wur.nl/upload_mm/e/f/b/6af2e2db-430e-4771-8f7d-6f5b974eab5e_final%20report%20ACT%202060%20juli%202018%20op%20website%20.pdf (accessed on 4 June 2019)

European Commission (2018) Final report of an audit carried out in Germany from 12 February 2018 to 21 February 2018 in order to evaluate member state activities to prevent tail-biting and avoid routine tail-docking of pigs; Ref. Ares (2018)4437429–29/08/2018; DG (SANTE) 2018–6445. Available online: http://ec.europa.eu/food/audits-analysis/audit_reports/details.cfm?rep_inspection_ref=2018-6445 (accessed on 4 June 2019)

Fraser D (1978) Observations on the behavioural development of suckling and early-weaned piglets during the first six weeks after birth. Anim Behav 26:22–30

Fraser D (2008) Understanding animal welfare. Acta Vet Scand 50:1–7

Fraser D, Weary DM, Pajor EA, Milligan BN (1997) A scientific conception of animal welfare that reflects ethical concerns. Anim Welf 6:187–205

Godyń D, Herbut E, Walczak J (2013) Infrared thermography as a method for evaluating the welfare of animals subjected to invasive procedures—a review. Ann Anim Sci 13:423–434

Guy JH, Meads ZA, Shiel RS, Edwards SA (2013) The effect of combining different environmental enrichment materials on enrichment use by growing pigs. Appl Anim Behav Sci 144:102–107

Hebb DO (1947) The effects of early experience on problem solving at maturity. Am Psychol 2:306–307

Hothersal B, Whistance L, Zedlacher Z, Algers B, Andersson E, Bracke M, Courboulay V, Ferrari P, Leeb C, Mullan S et al (2016) Standardising the assessment of environmental enrichment and tail-docking legal requirements for finishing pigs in Europe. Anim Welf 25:499–509

Keay JM, Singh J, Gaunt MC, Kaur T (2006) Fecal glucocorticoids and their metabolites as indicators of stress in various mammalian species: a literature review. J Zoo Wildl Med 37:234–244

Kelly HRC, Bruce JM, English RR, Fowler VR, Edwards SA (2000) Behaviour of 3 week weaned pigs in straw-flow, deep straw and flat desk housing systems. Appl Anim Behav Sci 68:269–280

Kersey DC, Dehnhard M (2014) The use of noninvasive and minimally invasive methods in endocrinology for threatened mammalian species conservation. Gen Comp Endocrinol 203:296–306

Kittawornat A, Zimmerman JJ (2010) Toward a better understanding of pig behavior and pig welfare. Anim Health Res Rev 12:25–32

Laloy F, Huet FR. Tail docking towards a national action plan of French pork production 2017. Available online: https://circabc.europa.eu/sd/a/362f9971-7f5c-4ece-93b2-560efc445e73/

National%20Plan%20on%20Tail%20Docking%20(FR)_LALOY%20F%20%26%20 HUET%20R_2017_EN.pdf (accessed on 4 June 2019)

Langen M, Kas MH, Staal WG, van Engeland H, Durston S (2011) The neurobiology of repetitive behavior: of mice. Neurosci Biobehav Rev 35:345–355

Mallapur A, Cheelam R (2002) Environmental influences on stereotypy and the activity budget of Indian leopards (*Panthera pardus*) in foru zoos in southern India. Zoo Biol 21:585–595

Martin JE, Ison SH, Baxter EM (2015) The influence of neonatal environment on piglet play behaviour and post-weaning social and cognitive development. Appl Anim Behav Sci 163:69–79

Mason G (1991) Stereotypies: a critical review. Anim Behav 41:1015–1037

Mason G, Rushen J (2006) Stereotypic behaviour in captive animals: fundamentals and implications for welfare and beyond. In: Stereotypic animal behaviour: fundamentals and applications to welfare, 2nd edn. CABI, Cambridge, pp 326–356

McCormick W. Recognising and assessing positive welfare: developing positive indicators for use in welfare assessment. In Proceedings of the Measuring Behavior, Utrecht, The Netherlands, 28–31 August 2012; Spink AJ, Grieco F, Krips OE, Loijens LWS, Noldus LPJJ, Zimmerman PH, Eds.; pp. 241–243

McGlone JJ (1986) Influence of resources on pig aggression and dominance. Behav Process 12:135–144

Mkwanazi MV, Ncobela CN, Kanengoni AT, Chimonyo M (2019) Effects of environmental enrichment on behaviour, physiology and performance of pigs—a review. Asian Australas J Anim Sci 32:1–13

Mormède P, Andanson S, Auperin B, Beerda B, Guemene D, Malmkvist J, Manteca X, Manteuffel G, Prunet P, Van Reenen CG et al (2007) Exploration of the hypothalamic-pituitary-adrenal function as a tool to evaluate animal welfare. Physiol Behav 92:317–339

Morris MC, Compas BE, Garber J (2012) Relations among posttraumatic stress disorder, comorbid major depression, and HPA function: a systematic review and meta-analysis. Clin Psychol Rev 32:301–315

Morrison RS, Johnston LJ, Hilbrands AM (2007) A note on the effects of two versus one feeder locations on the feeding behaviour and growth performance of pigs in a deep-litter, large group housing system. Appl Anim Behav Sci 107:157–161

Munsterhjelm C, Valros A, Heinonen M, Hälli O, Siljander-Rasi H, Peltoniemi OAT (2010) Environmental enrichment in early life affects cortisol patterns in growing pigs. Animal 4:242–249

Nannoni E, Sardi L, Vitali M, Trevisi E, Ferrari A, Ferri EM, Bacci M, Govoni N, Barbieri S, Martelli G (2018) Enrichment devices for undocked heavy pigs: E ects on animal welfare, blood parameters and production traits. Ital J Anim Sci 1–12

Nithianantharajah J, Hannan AJ (2006) Enriched environments, experience-dependent plasticity and disorders of the nervous system. Nat Rev Neurosci 7:697–709

Parratt CA, Chapman KJ, Turner C, Jones PH, Mendl MT, Miller BG (2006) The fighting behaviour of piglets mixed before and after weaning in the presence or absence of a sow. Appl Anim Behav Sci 101:54–67

Peeters E, Geers R (2006) Influence of provision of toys during transport on stress responses and meat quality of pigs. Anim Sci 82:591–595

Reimert I, Bolhuis JE, Kemp B, Rodenburg TB (2013) Indicators of positive and negative emotions and emotional contagion in pigs. Physiol Behav 109:42–50

Reynolds S, Lane SJ, Richards L (2010) Using animal models of enriched environments to inform research on sensory integration intervention for the rehabilitation of neurodevelopmental disorders. J Neurodev Disord 2:120–132

Salak-Johnson JL, Anderson DL, McGlone JJ (2004) Differential dose effects of central CRF and effects of CRF astressin on pig behavior. Physiol Behav 83:143–150

Schrøder-Petersen DL, Simonsen HB (2001) Tail biting in pigs. Vet J 162:196–210

Scott K, Chennells DJ, Campbell FM, Hunt B, Armstrong D, Taylor L, Gill BP, Edwards SA (2006) The welfare of finishing pigs in two contrasting housing systems: fully-slatted versus straw-bedded accommodation. Livest Sci 103:104–115

Scott K, Taylor L, Gill BP, Edwards SA (2007) Influence of different types of environmental enrichment on the behaviour of finishing pigs in two different housing systems: 2. Ratio of pigs to enrichment. Appl Anim Behav Sci 105:51–58

Spînka M, Newberry RC, Bekoff M (2001) Mammalian play: training for the un-expected. Q Rev Biol 76:141–168

Statham P, Green L, Mendl M (2011) A longitudinal study of the effects of providing straw at different stages of life on tail-biting and other behaviour in commercially housed pigs. Appl Anim Behav Sci 134:100–108

Stewart M, Webster J, Schaefer A, Cook N, Scott S (2005) Infrared thermography as a non-invasive tool to study animal welfare. Anim Welf 14:319–325

Taylor NR, Main DCJ, Mendl M, Edwards SA (2010) Tail-biting: a new perspective. Vet J 186:137–147

Telkänranta H, Bracke MBM, Valros A (2014) Fresh wood reduces tail and ear biting and increases exploratory behavior in finishing pigs. Appl Anim Behav Sci 161:51–59

The European Commission (2016) Commission Recommendation (EU) 2016/336 of 8 March 2016 on the application of Council Directive 2008/120/EC laying down minimum standards for the protection of pigs as regards measures to reduce the need for tail-docking. Off J Eur Union. Available online: https://eur-lex.europa.eu/eli/reco/2016/336/oj (accessed on 16 April 2019)

Turner CA, Lewis MH (2003) Environmental enrichment: effects on stereotyped behavior and neurotrophin levels. Physiol Behav 80:259–266

Turner SP, White IMS, Brotherstone S, Farnworth MJ, Knap PW, Penny P, Mendl M, Lawrence AB (2006) Heritability of post-mixing aggressiveness in grower-stage pigs and its relationship with production traits. Anim Sci 82:615–620

Van de Weerd HA, Docking CM, Day JEL, Avery PJ, Edwards SA (2003) A systematic approach towards developing environmental enrichment for pigs. Appl Anim Behav Sci 84:101–118

Van Dixhoorn IDE, Reimert I, Middelkoop J, Bolhuis JE, Wisselink HJ, Groot Koerkamp PWG, Kemp B, Stockhofe-Zurwieden N (2016) Enriched housing reduces disease susceptibility to co-infection with porcine reproductive and respiratory virus (PRRSV) and *Actinobacillus pleuropneumoniae* (*A. pleuropneumoniae*) in young pigs. PLoS One 11:e0161832

Wallgren T, Westin R, Gunnarsson S (2016) A survey of straw use and tail biting in Swedish pig farms rearing undocked pigs. Acta Vet Scand 58:84

Widowski TM, Torrey S, Bench CJ, Gonyou HW (2008) Development of ingestive behaviour and the relationship to belly nosing in early-weaned piglets. Appl Anim Behav Sci 110:109–127

Yang CH, Ko HL, Salazar LC, Llonch L, Manteca X, Camerlink I, Llonch P (2018) Pre-weaning environmental enrichment increases piglets' object play behaviour on a large scale commercial pig farm. Appl Anim Behav Sci 202:7–12

Zwicker B, Gygax L, Wechsler B, Weber R (2012) Influence of the accessibility of straw in racks on exploratory behaviour in finishing pigs. Livest Sci 148:67–73

Production of Biofuel from Pork Fat

Felix Uchenna Samuel and Jacob Oluwoye

Abstract

Biofuel production from pork fat is a promising alternative to traditional petroleum-based fuels. Pork fat, a waste product from the meat industry, can be converted into high-quality biodiesel through a process called transesterification. Researches have shown that biodiesel derived from pork lard has several advantages over diesel fuel, including a higher cetane number, lower sulfur content, and improved lubricity. The transesterification process involves reacting pork fat with a short-chain alcohol, typically methanol, in the presence of a catalyst such as sodium or potassium hydroxide. This produces fatty acid methyl esters (FAME), which are the main components of biodiesel, as well as glycerin as a byproduct. Optimizing the reaction conditions, such as temperature, catalyst concentration, and methanol-to-oil ratio, can improve the biodiesel yield and quality. Researchers have also explored the use of immobilized enzymes as catalysts for the transesterification of pork fat, which can simplify the purification process and allow for reuse of the catalyst. Additionally, the use of waste animal fats as a feedstock for biodiesel production can contribute to the sustainability and cost-effectiveness of the process. Overall, the production of biofuel from pork fat represents a viable and environmentally friendly alternative to traditional fossil fuels, with the potential to reduce greenhouse gas emissions and provide a valuable use for a waste product from the meat industry.

F. U. Samuel (✉)
Animal Science Program, Alabama Cooperative Extension, Alabama A&M University, Huntsville, AL, USA

J. Oluwoye
Departments of Urban and Regional Planning, Alabama A&M University, Huntsville, AL, USA

© The Author(s), under exclusive license to Springer Nature Singapore Pte Ltd. 2024
T. Rana, B. Soto-Blanco (eds.), *Good Practices and Principles in Pig Farming*, Livestock Diseases and Management,
https://doi.org/10.1007/978-981-97-4665-1_10

Keywords

Pig · Biofuel · Transesterification · Pork fat · Waste utilization · Biodiesel

10.1 Background

The global energy situation is going through a significant change due to concerns about climate change, energy security, and the decreasing availability of fossil fuels (Wang and Azam 2024). There is a growing interest in using biofuels, which are made from renewable organic materials, as a viable alternative to traditional petroleum-based fuels. Biofuels have the potential to reduce greenhouse gas emissions and dependence on limited resources. While biofuels have historically been made from crops like corn, sugarcane, and soybean, there is now a rising interest in using animal fats, specifically pork fat, as raw materials for biofuel production. Pork fat, which is a byproduct of the meat processing industry, has the potential to be a plentiful and underutilized resource for producing biofuels (Iyke 2024). By using pork fat for biofuel production, we not only find a sustainable solution for waste management but also create opportunities to diversify the sources of raw materials for biofuels and improve energy security (Skaggs et al. 2018).

10.2 Definition and Types of Biofuel

Biofuels are renewable fuels derived from organic materials, including plant biomass, animal fats, and microbial sources (Sungur 2024). They are produced through various biochemical and thermochemical processes, with the primary objective of providing an alternative to fossil fuels (Shweta et al. 2024). Biofuels can be categorized into three main types based on their feedstock sources and production methods (Fig. 10.1).

10.2.1 First-Generation Biofuels

First-generation biofuels are mainly manufactured using edible crops like corn, sugarcane, soybean, and palm oil. The primary types of first-generation biofuels are biodiesel and bioethanol. Biodiesel is typically made by converting vegetable oils or animal fats through a process called transesterification. On the other hand, bioethanol is produced by fermenting sugars or starches found in crops. These first-generation biofuels have been extensively employed as transportation fuels and are mixed with traditional petroleum fuels to lower greenhouse gas emissions and decrease reliance on fossil fuels (Dhir 2024).

10 Production of Biofuel from Pork Fat

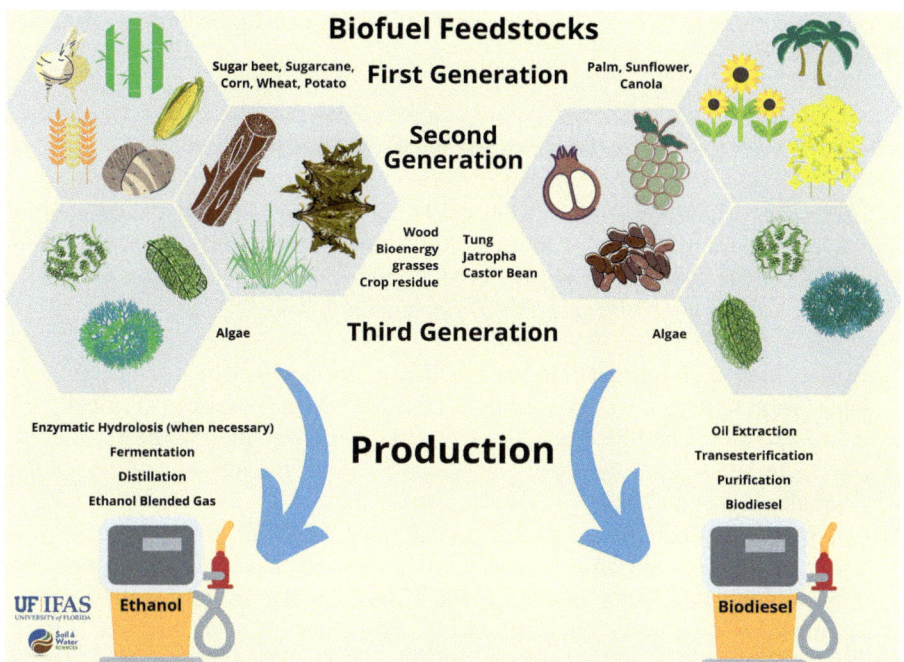

Fig. 10.1 Generations of biofuel production

10.2.2 Second-Generation Biofuels

Second-generation biofuels derive from non-food sources like lignocellulosic biomass, agricultural residues, and woody biomass. Processes like biochemical conversion, involving enzymatic hydrolysis and fermentation, as well as thermochemical conversion, including pyrolysis and gasification, are utilized to transform intricate carbohydrates and lignin into biofuels such as cellulosic ethanol, renewable diesel, and syngas. These advanced biofuels present various benefits compared to first-generation counterparts, such as enhanced sustainability, diminished reliance on food crops, and the ability to utilize waste materials (Sun 2024).

10.2.3 Third-Generation Biofuels

Third-generation biofuels originate from algae, microalgae, and other aquatic organisms. The production of these biofuels involves photosynthetic mechanisms, where microorganisms transform sunlight and carbon dioxide into lipids or hydrocarbons. Algal biofuels have attracted attention because of their significant productivity capacity, minimal land necessities, and capability to thrive in various settings, including non-arable land and wastewater treatment facilities. Nonetheless, there are obstacles related to the cultivation, harvesting, and extraction processes that must be overcome to fully exploit the potential of third-generation biofuels (Rabbani et al. 2024).

10.3 Sources of Feedstocks for Biofuel Production

The selection of raw materials is a critical factor in determining the sustainability, effectiveness, and feasibility of biofuel production. There have been investigations into numerous feedstocks for biofuel production, each with its own set of benefits and obstacles (Amal and Usman 2024). Typical feedstocks include:

Crop Residues: Agricultural leftovers such as corn stover, wheat straw, and rice husks represent plentiful and easily accessible biomass sources. Typically left unused post-harvest, these residues can undergo conversion into biofuels utilizing techniques like pyrolysis, gasification, and biochemical fermentation (Ramalingam et al. 2024).

Woody Biomass: Examples of woody biomass that can be utilized for biofuel production include forestry residues, sawdust, and wood chips. Woody biomass contains significant amounts of cellulose and lignin, which make it well-suited for thermochemical conversion methods like pyrolysis and gasification (Lugani et al. 2024).

Energy Crops: Specialized energy crops like switchgrass, miscanthus, and willow are deliberately grown for bioenergy applications. These crops boast elevated yields and are suitable for cultivation on marginal lands, thereby mitigating conflicts with food production (Ford et al. 2024).

Waste Materials: Biofuels can be produced from organic waste streams such as municipal solid waste, food waste, and wastewater sludge using methods like anaerobic digestion, fermentation, or thermochemical conversion (Saidu et al. 2024).

Animal Fats: Animal fats and greases, such as tallow, lard, and poultry fat, can undergo transformation into biodiesel via transesterification or hydrodeoxygenation procedures. Although less frequently utilized than vegetable oils, animal fats present potential benefits such as increased energy density and reduced production expenses (Jaiswal et al. 2024).

Microorganisms: Certain microorganisms, such as bacteria and yeast, can produce biofuels through fermentation processes. Microbial fermentation can convert a wide range of feedstocks, including sugars, starches, and lignocellulosic biomass, into bioethanol, biobutanol, and other bio-based chemicals (Sharma and Shankar 2024).

Algae: Microalgae and macroalgae are promising feedstocks for the production of third-generation biofuels such as biodiesel and bioethanol. Algae can be cultivated in a variety of environments, including freshwater ponds, saline water systems, and wastewater treatment facilities, offering scalability and flexibility in production (Jambo et al. 2016).

10.4 Production of Biofuel from Pork Fat

Pork fat, also known as lard, is a promising feedstock for the production of biodiesel, a renewable and sustainable liquid fuel. Biodiesel can be produced from various animal fats, including pork fat, through a process called transesterification (Praveena et al. 2024).

10.5 Characteristics of Pork Fat as a Biodiesel Feedstock

Pork fat is a suitable feedstock for biodiesel production due to its high content of saturated fatty acids, which are desirable for biodiesel properties. The fatty acid composition of pork fat typically includes:

Myristic acid (C14:0): 1.6%, Palmitic acid (C16:0): 25.4%, Stearic acid (C18:0): 13.6%, Oleic acid (C18:1): 44.0%, and Linoleic acid (C18:2): 10.0%.

The high proportion of saturated fatty acids, such as palmitic and stearic acids, contributes to the favorable cold flow properties of biodiesel produced from pork fat, making it suitable for use in colder climates. Additionally, the presence of monounsaturated fatty acids, like oleic acid, helps to improve the oxidative stability of the biodiesel (Praveena et al. 2024).

10.6 Importance and Benefits of Biofuel

The production and utilization of biofuels offer several important benefits (Camilo et al. 2024):

Renewable Energy Source: Biofuels originate from sustainable biomass sources that can be responsibly grown and renewed over time. In contrast to finite and non-renewable fossil fuels, biofuels provide a sustainable option for fulfilling energy needs while diminishing dependence on exhaustible resources.

Greenhouse Gas Emissions Reduction: Biofuels offer the potential to reduce greenhouse gas emissions by capturing carbon dioxide during biomass growth and compensating for emissions generated by fossil fuel combustion. In comparison to traditional petroleum fuels, biofuels generally exhibit reduced lifecycle carbon footprints, aiding in efforts to mitigate climate change and achieve emission reduction objectives outlined in global accords like the Paris Agreement. Furthermore, certain biofuel production methods, such as anaerobic digestion of organic waste and utilization of agricultural residues, may yield negative net emissions by preventing methane emissions from decomposition and delivering carbon sequestration advantages.

Energy Security and Independence: Biofuels provide a chance to improve energy security by diversifying energy sources and decreasing reliance on imported fossil fuels. By utilizing locally sourced biomass materials, nations can enhance their energy self-sufficiency, reduce exposure to geopolitical risks, and stimulate rural economic growth through agricultural diversification and the creation of jobs in the bioenergy industry. Moreover, biofuel production can help mitigate fluctuations in energy prices by offering stable and predictable fuel supplies, particularly in regions with ample biomass resources.

Rural Development and Poverty Alleviation: The production of biofuels from agricultural feedstocks has the potential to stimulate rural economies and alleviate poverty by creating employment opportunities, supporting smallholder farmers, and fostering value-added agricultural activities. Biofuel production chains often involve multiple stakeholders, including farmers, processors, distributors, and retailers, thereby generating income and enhancing livelihoods across the agricultural value chain. Moreover, investments in bioenergy infrastructure and technology transfer initiatives can empower rural communities with access to modern energy services, improving living standards and reducing energy poverty in remote areas.

10.7 Challenges and Limitations of Biofuels

The production of biofuels is not without limitations and challenges (Ismaeel et al. 2024). These are outlined below:

Feedstock Availability and Competition: The availability and cost of feedstocks can vary depending on factors such as crop yields, land availability, and competing land uses (e.g., food production, conservation).

Land Use Change: The expansion of biofuel feedstock cultivation can lead to land use change, deforestation, and biodiversity loss, particularly if not managed sustainably.

Technological Maturity: Some biofuel conversion technologies are still in the early stages of development and may face technical, economic, and scalability challenges.

Energy Balance and Emissions: The net energy balance and environmental performance of biofuels can vary depending on factors such as feedstock choice, production methods, and lifecycle analysis assumptions.

Policy and Regulatory Uncertainty: The biofuels industry is subject to evolving policy frameworks, incentives, and regulations, which can impact investment decisions and market dynamics.

10.8 Pork Industry and Fat Byproducts

10.8.1 Global Pork Production

The pork industry plays a significant role in global agriculture and food production, providing a major source of protein and essential nutrients for human consumption. Pork is one of the most widely consumed meats worldwide, with high demand in regions such as Asia, Europe, and North America (Drewnowski 2024).

Global pork production has steadily increased over the past decades, driven by population growth, rising incomes, and changing dietary preferences. According to the Food and Agriculture Organization (FAO), pork production reached over 110 million metric tons in 2020, with China, the European Union, and the United States accounting for the majority of global output. Pork production systems vary widely across countries and regions, ranging from extensive outdoor systems to intensive indoor operations. Common pig breeds include Duroc, Landrace, Yorkshire, and Hampshire, each selected for specific traits such as growth rate, feed efficiency, and meat quality (Yu et al. 2024).

10.8.2 Composition and Properties of Pork Fat

Pork fat, also known as lard or pig fat, is a valuable co-product of pork processing with diverse applications in food, pharmaceutical, and industrial sectors. Pork fat is primarily composed of triglycerides, fatty acids, and minor components such as cholesterol, phospholipids, and vitamins. The composition of pork fat can vary depending on factors such as pig genetics, diet, age, and production methods. Common fatty acids found in pork fat include saturated fatty acids (e.g., palmitic acid, stearic acid), monounsaturated fatty acids (e.g., oleic acid), and polyunsaturated fatty acids (e.g., linoleic acid, linolenic acid) (Vongsawasdi and Noomhorm 2014).

Pork fat is characterized by its solid consistency at room temperature, high energy content (approximately 9 kcal/g), and unique flavor profile. These properties make pork fat suitable for various culinary applications, including frying, baking, and flavor enhancement in processed foods (Kesarwani et al. 2024) (Fig. 10.2).

10.8.3 Byproducts of Pork Processing

The pork industry generates a significant amount of byproducts during the processing of pork carcasses, including both edible and inedible parts. These byproducts are typically separated into various categories based on their intended use and value-added potential (Fig. 10.3). Common byproducts of pork processing include:

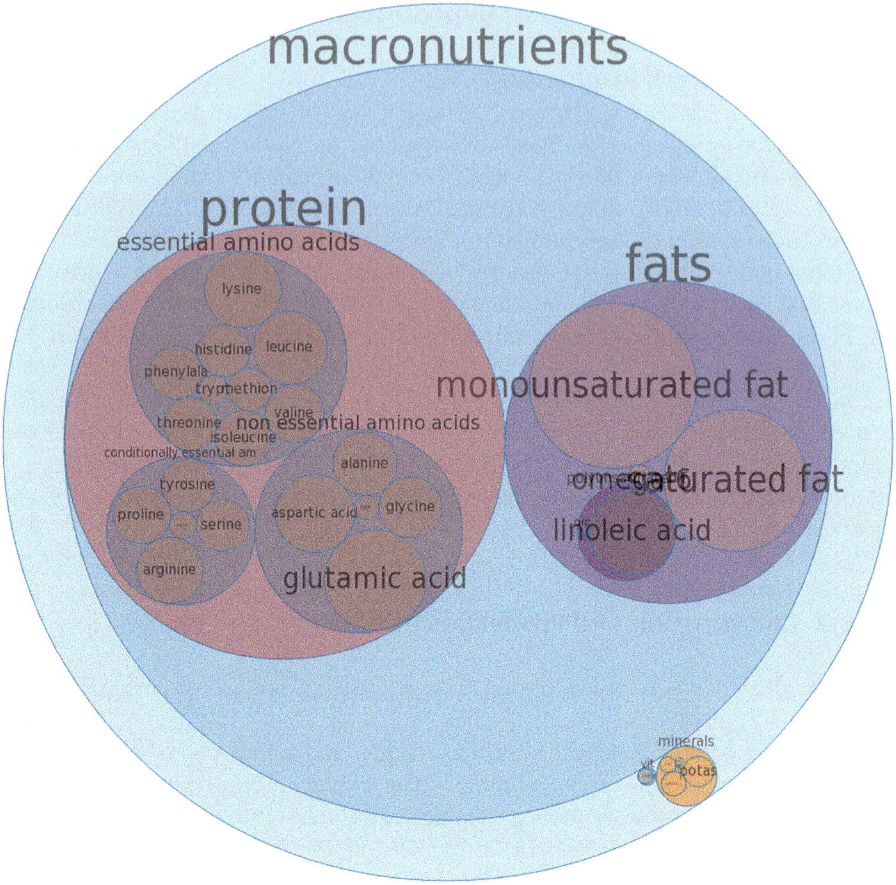

Fig. 10.2 Composition of pork fat

Fig. 10.3 Pigs by product

Meat Cuts: High-value cuts such as pork loin, pork chops, and pork tenderloin are typically sold fresh or processed into cured, smoked, or marinated products. These cuts are prized for their tenderness, flavor, and versatility in cooking.

Offal: Offal refers to the internal organs and entrails of slaughtered animals, including the heart, liver, kidneys, lungs, and intestines. While some offal is consumed as food, either fresh or processed, others are utilized for pet food, animal feed, or pharmaceutical purposes.

Blood: Blood is collected during the slaughter process and can be processed into blood meal or blood plasma for use in animal feed formulations. Blood-derived products are rich in protein, amino acids, and minerals, making them valuable nutritional supplements.

Bones: Pork bones contain collagen, minerals, and trace elements that can be extracted through processes such as rendering, boiling, or enzymatic hydrolysis. Bone-derived products, such as gelatin, bone meal, and bone broth, are used in food, pharmaceutical, and industrial applications.

Fat: Pork fat is separated from the carcass during the butchering process and can be rendered into lard or used for other purposes such as cooking oil, biodiesel feedstock, or industrial lubricants. Pork fat is prized for its flavor-enhancing properties and high energy content.

Skin: Pork skin, also known as pork rind or pork crackling, is a popular snack food in many cultures. It can also be processed into gelatin, collagen peptides, or animal feed additives. Pork skin is valued for its crispy texture and savory flavor.

Connective Tissue: Connective tissue, including tendons, ligaments, and cartilage, is often processed into gelatin or collagen hydrolysate for use in food, cosmetics, and pharmaceutical products. These collagen-based ingredients are valued for their gelling, thickening, and stabilizing properties.

10.9 Biochemical Conversion of Pork Fat to Biofuels

10.9.1 Transesterification Process

Transesterification is the most common method used to convert pork fat into biodiesel, a renewable diesel fuel that can be used in diesel engines with little or no modification. The transesterification process involves the reaction of pork fat triglycerides with alcohol (typically methanol or ethanol) in the presence of a catalyst (usually sodium hydroxide or potassium hydroxide) (Singh et al. 2024). During transesterification, the alcohol reacts with the ester groups in the pork fat triglycerides, resulting in the formation of fatty acid alkyl esters (biodiesel) and glycerol as a byproduct. The reaction is typically carried out at elevated temperatures (50–60 °C) and atmospheric pressure, although variations in reaction conditions may be employed to optimize yield and efficiency. Once the transesterification reaction is complete, the resulting biodiesel-glycerol mixture undergoes separation to recover the biodiesel product. This is typically achieved through processes such as gravity settling, centrifugation, or membrane filtration, which separate the biodiesel phase from the glycerol phase (Singh et al. 2024).

10.9.1.1 Advantages of Transesterification

Versatility: Transesterification can utilize a wide range of alcohol sources (e.g., methanol, ethanol) and catalysts (e.g., sodium hydroxide, potassium hydroxide), offering flexibility in process design and optimization.

High Conversion Efficiency: Transesterification can achieve high conversion rates (>95%) of pork fat triglycerides into biodiesel under appropriate reaction conditions.

Mild Reaction Conditions: Transesterification can be carried out at relatively mild reaction temperatures and pressures, reducing energy requirements and operating costs.

Biodiesel Quality: Biodiesel produced via transesterification meets established fuel quality standards (e.g., ASTM D6751, EN 14214) for properties such as cetane number, kinematic viscosity, and sulfur content.

10.9.1.2 Limitations and Challenges of Transesterification

Feedstock Quality: The quality of pork fat feedstock can impact the efficiency and yield of transesterification reactions, with impurities such as free fatty acids, moisture, and solid contaminants affecting catalyst performance and product quality.

Catalyst Handling: The handling and disposal of catalysts used in transesterification processes (e.g., sodium hydroxide, potassium hydroxide) require careful management to minimize environmental impact and ensure worker safety.

Glycerol Disposal: The glycerol byproduct generated during transesterification must be separated, purified, and disposed of properly to prevent contamination and environmental pollution. Glycerol can be further processed into value-added products or utilized in other industrial applications to improve overall process economics (Fig. 10.4).

10.9.2 Hydrogenation Process

Hydrogenation presents an alternative approach to transform pork fat into renewable diesel fuel, termed hydrotreated renewable diesel (HRD) or green diesel. In contrast to transesterification, where triglycerides undergo alcoholysis, hydrogenation employs hydrogen gas (H_2) to chemically diminish the double bonds in fatty acids derived from pork fat, yielding saturated hydrocarbons that closely resemble petroleum diesel (Barla et al. 2024) (Fig. 10.5).

10.9.2.1 Steps of Hydrogenation Process

Pretreatment: Pork fat feedstock is pretreated to remove impurities such as moisture, free fatty acids, and solid contaminants that could deactivate catalysts or cause equipment fouling.

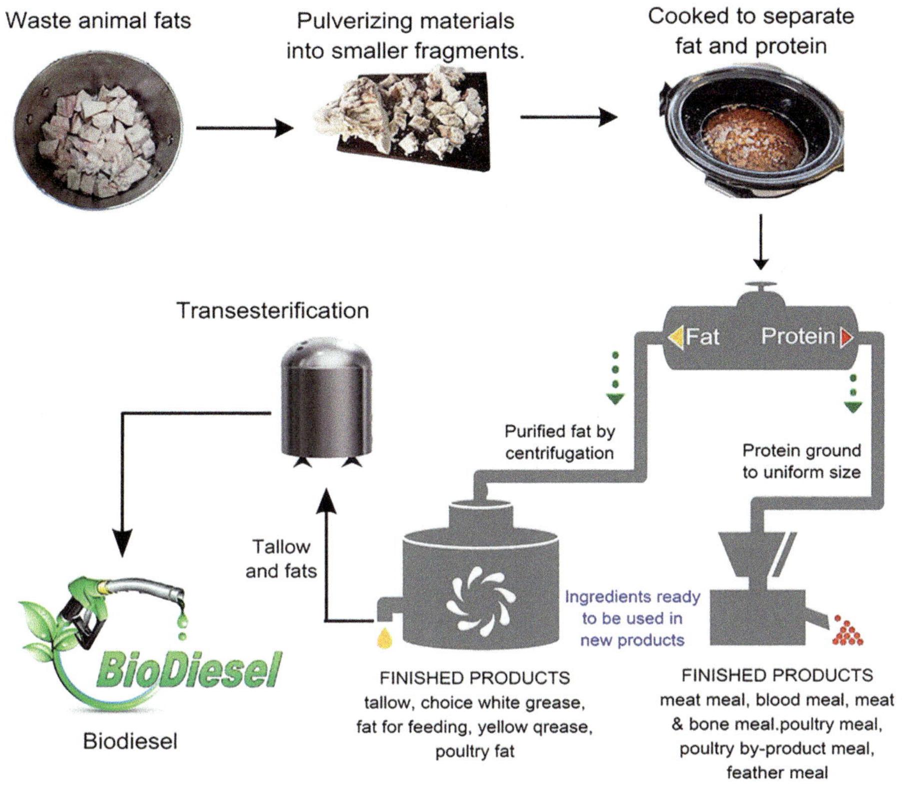

10 Production of Biofuel from Pork Fat

Fig. 10.4 Transesterification of animal fat

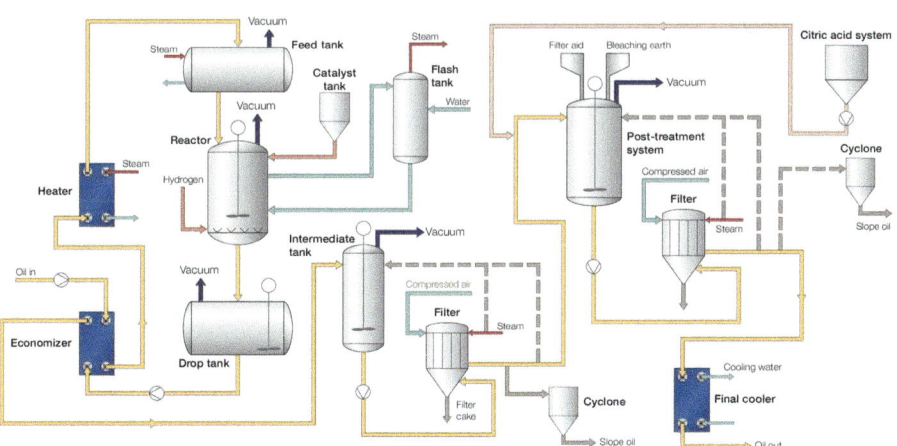

Fig. 10.5 Hydrogenation process

Hydrogenation Reaction: The pretreated pork fat is mixed with hydrogen gas and passed over a heterogeneous catalyst (e.g., nickel, palladium, platinum) in a fixed-bed reactor at elevated temperatures (200–300 °C) and high pressures (10–100 bar).

Product Separation: The hydrogenation reaction produces a mixture of hydrocarbons, including diesel-range paraffins and cyclic compounds. The product mixture is then separated and purified through processes such as distillation, solvent extraction, or hydrotreating to obtain the desired HRD product.

10.9.2.2 Advantages of Hydrogenation Process

High Yield and Selectivity: Hydrogenation can achieve high conversion rates (>90%) of pork fat triglycerides into diesel-range hydrocarbons with minimal formation of unwanted byproducts.

Superior Fuel Properties: HRD produced via hydrogenation exhibits properties similar to petroleum diesel, including high cetane number, low sulfur content, and excellent cold flow properties, making it compatible with existing diesel engines and fuel distribution infrastructure.

Reduced Emissions: HRD has lower emissions of particulate matter, nitrogen oxides, and sulfur oxides compared to conventional diesel fuel, contributing to air quality improvement and environmental sustainability.

10.9.2.3 Problems of Hydrogenation Process

Catalyst Deactivation: The presence of impurities in pork fat feedstock can lead to catalyst deactivation, reducing hydrogenation efficiency and necessitating frequent catalyst regeneration or replacement.

Hydrogen Requirements: Hydrogenation processes require large quantities of hydrogen gas, which must be generated from renewable sources or derived from fossil fuels through steam reforming or other processes. The sustainability and environmental impact of hydrogen production methods should be considered.

Process Complexity: Hydrogenation processes involve high temperatures, pressures, and specialized catalysts, making them more complex and capital-intensive compared to transesterification. Process optimization and reactor design are critical to maximizing efficiency and yield.

Product Distribution: The distribution of hydrocarbon products in HRD depends on factors such as reaction conditions, catalyst type, and feedstock composition. Achieving the desired diesel-range hydrocarbons requires careful control of reaction parameters and catalyst performance.

10.9.3 Pyrolysis and Gasification

Pyrolysis and gasification are thermochemical conversion processes that can be used to convert pork fat into syngas, a mixture of hydrogen, carbon monoxide, and methane, which can be further processed into biofuels such as ethanol, methanol, or Fischer-Tropsch diesel. Pyrolysis involves the thermal decomposition of organic

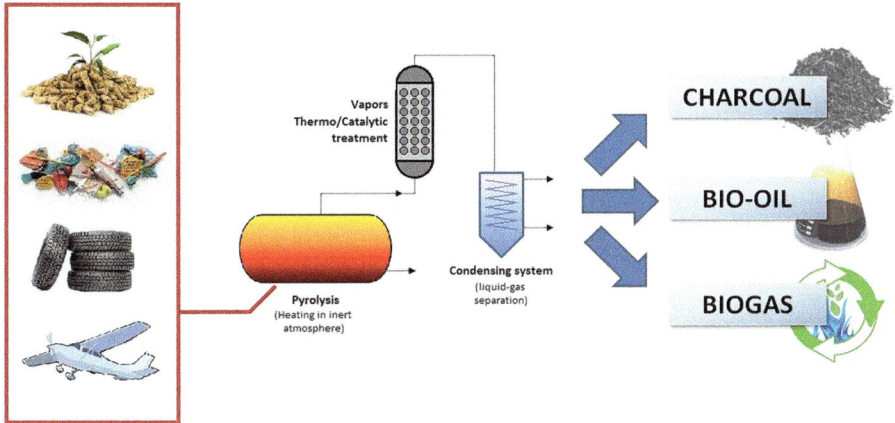

Fig. 10.6 Pyrolysis and gasification

materials in the absence of oxygen, resulting in the production of bio-oil, biochar, and syngas. Pork fat feedstock is heated to high temperatures (>400 °C) in a reactor, causing the breakdown of triglycerides into smaller molecules. The resulting bio-oil can be upgraded through processes such as hydrodeoxygenation to remove oxygen and improve fuel properties (Saidu et al. 2024) (Fig. 10.6).

Gasification is a similar process to pyrolysis but operates at higher temperatures (>700 °C) and introduces a limited amount of oxygen or steam to facilitate partial oxidation of the feedstock. Pork fat feedstock undergoes thermal decomposition and gasification reactions, producing a syngas rich in hydrogen and carbon monoxide. The syngas can be cleaned, cooled, and processed into biofuels using catalytic conversion technologies.

10.9.3.1 Advantages of Pyrolysis and Gasification
This offers several advantages for the production of biofuels from pork fat, including:

Feedstock Flexibility: Pyrolysis and gasification can utilize a wide range of feedstocks, including pork fat, agricultural residues, woody biomass, and waste materials, offering flexibility in feedstock selection and availability.
Syngas Production: Pyrolysis and gasification produce syngas as an intermediate product, which can be further processed into various biofuels or used for heat and power generation in combined heat and power (CHP) systems.
Waste Reduction: Pyrolysis and gasification can process organic waste streams, including pork fat byproducts, into valuable biofuels and bio-based products, reducing waste disposal costs and environmental impacts.

10.9.3.2 Challenges and Limitations of Pyrolysis and Gasification
These include the following:

Complex Process Design: Pyrolysis and gasification systems require sophisticated equipment and control systems to manage high temperatures, pressures, and reaction kinetics. Process integration and optimization are critical to maximizing efficiency and product yields.

Product Yield and Quality: The composition and properties of bio-oil and syngas produced from pyrolysis and gasification can vary depending on feedstock characteristics, reactor design, and operating conditions. Achieving consistent product quality and yield requires careful process control and optimization.

Biomass Ash and Tar Formation: Pyrolysis and gasification can produce ash and tar as byproducts, which can cause fouling and corrosion in reactor systems and downstream equipment. Tar removal and management are essential for maintaining system performance and reliability.

10.9.4 Microbial Conversion

Microbial conversion is a biological process that utilizes microorganisms such as bacteria, yeast, and fungi to convert pork fat into biofuels through fermentation, anaerobic digestion, or enzymatic hydrolysis (Nwokolo and Enebe 2024).

Fermentation involves the anaerobic conversion of sugars or organic compounds into biofuels such as ethanol, biobutanol, or microbial oils by microbial fermentation. Pork fat triglycerides can be hydrolyzed into fatty acids and glycerol using lipase enzymes, followed by fermentation of the fatty acids into biofuels by yeast or bacteria. Fermentation processes can operate at mild conditions and are well-suited for converting high-fat feedstocks like pork fat into biofuels.

Anaerobic digestion is a biological process that converts organic materials into biogas (a mixture of methane and carbon dioxide) through microbial decomposition in the absence of oxygen. Pork fat feedstock is mixed with anaerobic microorganisms in a digester, where it undergoes hydrolysis, acidogenesis, acetogenesis, and methanogenesis stages to produce biogas. Biogas can be upgraded to biomethane and used as a renewable natural gas or transportation fuel.

Enzymatic hydrolysis involves the use of enzymes such as lipases, esterases, and proteases to break down pork fat triglycerides into fatty acids and glycerol. The resulting fatty acids can be further converted into biofuels such as biodiesel or renewable diesel using microbial fermentation or chemical processes. Enzymatic hydrolysis offers mild reaction conditions, high selectivity, and compatibility with a wide range of feedstocks.

10.9.4.1 Advantages of Microbial Conversion
These include the following:

Mild Reaction Conditions: Microbial conversion processes operate at mild temperatures and pressures, reducing energy requirements and environmental impacts compared to thermochemical conversion methods.

10 Production of Biofuel from Pork Fat

Feedstock Flexibility: Microbial conversion can utilize a wide range of feedstocks, including pork fat, waste oils, and greases, offering flexibility in feedstock selection and availability.

Biodegradation of Impurities: Microbial conversion processes can biodegrade impurities such as free fatty acids, glycerides, and contaminants present in pork fat feedstock, reducing the need for pretreatment and purification steps.

Renewable Energy Production: Microbial conversion processes produce biofuels such as ethanol, biogas, and biodiesel from renewable feedstocks, contributing to energy security and sustainability.

10.9.4.2 Challenges and Limitations of Microbial Conversion
These include the following:

Process Efficiency: Microbial conversion processes may have lower conversion efficiencies and yields compared to thermochemical conversion methods, particularly for high-fat feedstocks like pork fat.

Microbial Activity: Microbial conversion processes depend on the activity and stability of microorganisms, which can be influenced by factors such as temperature, pH, nutrient availability, and substrate composition.

Process Scale-Up: Scaling up microbial conversion processes from laboratory or pilot scale to commercial production can be challenging due to issues such as reactor design, process optimization, and microbial cultivation.

10.10 Technological Innovations in Pork Fat Biofuel Production

10.10.1 Catalyst Development

Catalysts play a critical role in biochemical conversion processes, facilitating the transformation of pork fat into biofuels with enhanced efficiency and selectivity (e Melo et al. 2024). Ongoing research efforts are focused on developing novel catalyst materials, improving catalyst performance, and optimizing reaction kinetics for various conversion pathways.

Transesterification processes, catalysts such as alkaline metal hydroxides (e.g., sodium hydroxide, potassium hydroxide) are commonly used to facilitate the reaction between pork fat triglycerides and alcohol. Recent advancements in catalyst development include the use of heterogeneous catalysts such as solid acids, metal oxides, and supported nanoparticles, which offer improved catalytic activity, stability, and recyclability compared to traditional homogeneous catalysts.

Hydrogenation processes also rely on catalysts to convert pork fat fatty acids into diesel-range hydrocarbons. Catalyst development efforts focus on optimizing catalyst composition, structure, and morphology to enhance hydrogenation activity, selectivity, and resistance to deactivation. Bimetallic catalysts, metal-promoted supports, and nanostructured materials show promise in improving hydrogenation efficiency and reducing catalyst costs.

Pyrolysis and gasification processes, catalysts are used to control reaction pathways, increase syngas yields, and minimize undesirable byproducts such as tar and char. Catalysts such as zeolites, transition metal catalysts, and alkali-metal salts are being investigated for their ability to enhance biomass decomposition, gasification kinetics, and syngas quality. Integrated catalytic reactors and multifunctional catalyst systems are also under development to improve process efficiency and flexibility.

Microbial conversion processes rely on enzymes, microorganisms, and cofactors to catalyze biochemical reactions involved in pork fat degradation and biofuel production. Research in this area focuses on enzyme engineering, strain selection, and metabolic engineering to optimize microbial performance, substrate utilization, and product yields. Immobilized enzymes, genetically modified microbes, and synthetic biology approaches are being explored to enhance process robustness, productivity, and scalability.

10.10.2 Process Optimization

Process optimization plays a crucial role in maximizing the efficiency, yield, and sustainability of pork fat biofuel production processes (Gonçalves et al. 2020). Advanced modeling and simulation techniques, coupled with experimental validation and optimization strategies, are employed to identify optimal process conditions, design parameters, and operating strategies.

Transesterification processes, process optimization aims to minimize reaction time, energy consumption, and waste generation while maximizing biodiesel yield and quality. Factors such as temperature, pressure, alcohol-to-oil ratio, catalyst concentration, and agitation speed are optimized to achieve the desired conversion efficiency and product characteristics. Advanced reactor designs, such as continuous-flow reactors, oscillatory flow reactors, and microreactors, offer improved mass and heat transfer rates, reaction kinetics, and scalability compared to traditional batch reactors.

Hydrogenation processes require careful optimization of reaction parameters, catalyst properties, and hydrogen utilization to achieve high conversion rates and diesel-range hydrocarbon yields. Process variables such as temperature, pressure, hydrogen-to-feed ratio, residence time, and catalyst loading are optimized to balance hydrogenation activity, selectivity, and catalyst stability. Integrated process configurations, such as staged reactors, cascade systems, and membrane reactors, enable efficient hydrogenation of pork fat fatty acids while minimizing energy consumption, equipment costs, and environmental impacts.

Pyrolysis and gasification processes undergo optimization to maximize syngas yield, quality, and composition while minimizing energy losses, tar formation, and environmental emissions. Process parameters such as temperature, pressure, residence time, feedstock particle size, and gasification agent composition are optimized to enhance biomass conversion efficiency, syngas heating value, and process robustness. Advanced reactor designs, such as fluidized bed reactors, entrained flow reactors, and plasma gasifiers, offer improved thermal management, heat transfer, and product selectivity compared to conventional fixed-bed reactors.

Microbial conversion processes are optimized to enhance microbial growth, substrate utilization, and biofuel production while minimizing fermentation time, nutrient requirements, and contamination risks. Operating conditions such as pH, temperature, agitation, aeration, and substrate concentration are optimized to support microbial metabolism, enzyme activity, and product formation. In situ monitoring and control strategies, such as online sensors, real-time analytics, and feedback control loops, enable dynamic process optimization and adaptive operation in response to changing conditions.

10.10.3 Integration with Existing Infrastructure

Integration with existing infrastructure is crucial for the commercialization and adoption of pork fat biofuel production technologies, leveraging synergies, minimizing capital costs, and enhancing market competitiveness (Shahzad and Cheema 2024). Integration strategies focus on co-location, co-processing, and co-production opportunities within existing industrial sectors and supply chains.

Transesterification processes, integration with existing biodiesel production facilities, refineries, and chemical plants enables efficient utilization of infrastructure, utilities, and resources. Co-location of pork fat processing facilities with soybean crushing plants, vegetable oil refineries, or animal rendering facilities allows for shared infrastructure, feedstock logistics, and product distribution networks. Co-processing of pork fat feedstocks with other fats, oils, and greases (FOG) can improve feedstock flexibility and utilization rates, maximizing overall biodiesel production capacity and economies of scale. Additionally, integration with existing distribution networks, blending terminals, and fueling stations facilitates market penetration and supply chain logistics for distributing pork fat-derived biodiesel to end-users.

Hydrogenation processes can be integrated with existing petroleum refining infrastructure and hydroprocessing units to produce hydrotreated renewable diesel (HRD) from pork fat feedstocks. Co-processing of pork fat with petroleum feedstocks in existing hydrotreaters, hydrotreaters, and hydrocrackers allows for flexible production scheduling, feedstock blending, and product optimization. Integration with existing transportation fuel distribution networks, storage facilities, and retail outlets enables seamless market access and consumer acceptance of HRD as a drop-in replacement for petroleum diesel.

Pyrolysis and gasification processes can be integrated with existing biomass processing facilities, waste management facilities, and energy generation systems to produce biofuels from pork fat and other biomass feedstocks. Co-location of pyrolysis/gasification plants with agricultural processing facilities, municipal solid waste (MSW) treatment plants, or wastewater treatment facilities offers opportunities for feedstock synergies, waste valorization, and resource recovery. Integration with existing cogeneration plants, power plants, or industrial facilities allows for the co-production of biofuels, heat, power, and value-added products, enhancing overall process economics and sustainability.

Microbial conversion processes can be integrated with existing biorefineries, ethanol plants, and fermentation facilities to produce biofuels from pork fat and other organic feedstocks. Co-location of microbial conversion facilities with ethanol distilleries, biogas plants, or biochemical production units allows for shared infrastructure, utilities, and fermentation equipment. Integration with existing wastewater treatment plants, agro-industrial complexes, or food processing facilities offers opportunities for utilizing waste streams, byproducts, and co-products as feedstocks for microbial conversion, enhancing resource efficiency and circular economy principles.

10.10.4 Sustainability and Lifecycle Analysis

Sustainability and lifecycle analysis are essential tools for evaluating the environmental, economic, and social impacts of pork fat biofuel production processes across the entire value chain, from feedstock cultivation and processing to fuel distribution and end-use combustion (Lin and Lu 2021). Comprehensive lifecycle assessments (LCAs) quantify key performance indicators, such as greenhouse gas emissions, energy efficiency, resource consumption, land use change, and socio-economic benefits, to inform decision-making and support continuous improvement efforts.

Transesterification processes, sustainability considerations include feedstock sourcing, feedstock-to-fuel conversion efficiency, and lifecycle emissions reductions compared to fossil diesel. LCAs assess the environmental impacts of pork fat production, transportation, processing, and biodiesel distribution, accounting for factors such as land use change, fertilizer use, pesticide application, water consumption, and soil erosion. Optimization strategies, such as co-location of biodiesel facilities with animal rendering plants or waste oil collection centers, can enhance feedstock logistics, minimize transportation emissions, and improve overall process sustainability.

Hydrogenation processes undergo sustainability assessments to evaluate the carbon intensity, energy efficiency, and environmental performance of hydrotreated renewable diesel (HRD) compared to petroleum diesel. LCAs quantify the greenhouse gas emissions reductions, air pollutant emissions, water usage, and energy consumption associated with pork fat hydrogenation and HRD production. Integration with existing petroleum refining infrastructure, renewable energy sources, and carbon capture technologies can further enhance the sustainability credentials of HRD and support regulatory compliance with low carbon fuel standards (LCFS) and renewable fuel mandates.

Pyrolysis and gasification processes are subject to sustainability analysis to assess the net environmental benefits, resource efficiency, and socio-economic implications of biofuel production from pork fat and other biomass feedstocks.

LCAs evaluate the lifecycle greenhouse gas emissions, energy return on investment (EROI), waste generation, and ecosystem impacts associated with pyrolysis/gasification pathways, considering factors such as feedstock availability, process efficiency, and end-use applications. Co-production of biochar, bio-oil, and syngas offers opportunities for carbon sequestration, soil amendment, and renewable energy generation, contributing to circular economy principles and sustainable development goals.

Microbial conversion processes are evaluated for their sustainability performance in terms of feedstock utilization, waste reduction, and biofuel production efficiency. LCAs assess the environmental impacts, resource requirements, and ecosystem services associated with microbial fermentation, enzymatic hydrolysis, and anaerobic digestion pathways, considering factors such as feedstock diversity, process flexibility, and co-product valorization. Integration with biogas upgrading systems, nutrient recovery technologies, and decentralized biorefinery concepts can enhance the overall sustainability and resilience of microbial conversion systems, supporting climate change mitigation and rural development objectives.

10.11 Market Outlook and Future Directions

10.11.1 Market Trends and Drivers

The global biofuels market is experiencing significant growth and transformation, driven by a combination of regulatory mandates, climate policies, technological innovations, and market dynamics (Morone et al. 2023). Key trends and drivers shaping the market outlook for pork fat biofuels include:

Renewable Fuel Standards (RFS) and Low Carbon Fuel Standards (LCFS): Regulatory mandates and incentive programs in major markets such as the United States, European Union, and Brazil require blending of renewable fuels, including biodiesel, renewable diesel, and advanced biofuels, into transportation fuel supply chains to reduce greenhouse gas emissions and promote energy security.

Carbon Pricing and Emissions Trading: Increasing carbon pricing mechanisms, emissions trading schemes, and carbon tax initiatives create financial incentives for reducing carbon intensity and transitioning to low-carbon alternatives, including biofuels derived from pork fat and other renewable feedstocks.

Energy Security and Resilience: Concerns about energy security, geopolitical risks, and supply chain vulnerabilities are driving investments in domestic biofuel production, diversification of feedstock sources, and development of alternative fuel technologies to reduce dependence on imported oil and mitigate price volatility in global energy markets.

Sustainable Development Goals (SDGs): Growing awareness of environmental sustainability, social equity, and economic development goals outlined in the United Nations Sustainable Development Goals (SDGs) is driving demand for sustainable biofuels that contribute to climate change mitigation, rural livelihoods, and inclusive growth.

Technological Innovation and Cost Reduction: Advances in biofuel production technologies, process efficiencies, and feedstock utilization are reducing production costs, improving product quality, and enhancing competitiveness of pork fat biofuels relative to conventional fossil fuels.

Consumer Preferences and Corporate Sustainability Goals: Increasing consumer demand for eco-friendly products, corporate sustainability commitments, and supply chain transparency are driving adoption of biofuels with lower carbon foot prints and environmental credentials. Biofuels derived from pork fat offer a renewable, low-carbon alternative to petroleum-based fuels, aligning with consumer preferences for sustainable energy solutions and corporate sustainability goals for reducing carbon emissions and supporting renewable energy sources.

10.11.2 Growth Opportunities and Challenges

The pork fat biofuels market presents several growth opportunities and challenges that will shape its development and expansion in the coming years (Gbadeyan et al. 2024):

Feedstock Availability and Cost: The availability and cost of pork fat feedstock are key determinants of biofuel production economics and competitiveness. Access to reliable, affordable sources of pork fat, such as rendered animal fats from meat processing facilities, slaughterhouses, and food service establishments, is essential for maintaining stable production volumes and pricing.

Technology Maturation and Scale-Up: Continued advancements in biofuel production technologies, process optimization, and scale-up capabilities are needed to overcome technical challenges, reduce production costs, and improve product quality and consistency. Research and development efforts focused on catalyst development, reactor design, and process integration will drive innovation and commercialization of pork fat biofuel technologies.

Regulatory Frameworks and Policy Support: Supportive regulatory frameworks, policy incentives, and market mechanisms are critical for fostering investment, stimulating demand, and creating a level playing field for biofuels in the energy market. Clear, stable policy signals, such as renewable fuel mandates, tax incentives, and carbon pricing mechanisms, provide certainty and confidence for biofuel producers, investors, and stakeholders.

Infrastructure and Market Access: Infrastructure constraints, including storage, transportation, blending, and distribution facilities, pose challenges for integrating pork fat biofuels into existing fuel supply chains and accessing end-user markets. Investment in infrastructure upgrades, retrofits, and expansions is needed to accommodate biofuel blending, storage, and distribution requirements and ensure reliable supply to consumers.

Sustainability and Certification: Addressing sustainability concerns, including land use change, deforestation, biodiversity conservation, and social impacts, is essential for gaining public acceptance, regulatory approval, and market access for pork fat biofuels. Certification schemes, such as the Roundtable on Sustainable Biomaterials (RSB), the International Sustainability and Carbon Certification (ISCC), and the Sustainable Aviation Fuel (SAF) certification program, provide frameworks for demonstrating compliance with sustainability criteria and verifying the environmental and social sustainability of biofuel production.

Market Dynamics and Competitiveness: Market dynamics, including energy prices, feedstock costs, policy support, technological innovation, and consumer preferences, influence the competitiveness and profitability of pork fat biofuels relative to conventional fossil fuels. Price volatility, supply chain disruptions, and competition from other renewable fuels, such as soybean biodiesel, palm oil biodiesel, and waste-based biofuels, present challenges for market penetration and growth.

International Trade and Market Access: Global trade dynamics, trade agreements, tariffs, and import/export restrictions impact the competitiveness and market access of pork fat biofuels in international markets. Access to export markets, import tariffs, and trade barriers can affect the profitability and viability of biofuel production and trade, influencing investment decisions and supply chain strategies.

10.12 Emerging Trends and Future Directions

10.12.1 Circular Economy Integration

The integration of pork fat biofuel production into circular economy frameworks presents opportunities to enhance resource efficiency, waste valorization, and sustainability. Circular economy principles emphasize closing material loops, minimizing waste generation, and maximizing resource utilization through strategies such as reuse, recycling, and regeneration (Ezeorba et al. 2024). In the context of pork fat biofuels, circular economy integration involves:

Valorization of Byproducts: Byproducts generated from pork fat biofuel production, such as glycerol, fatty acids, and residual biomass, can be valorized through conversion into value-added products, including biochemicals, bioplastics, animal feed additives, and soil amendments. Integrated biorefinery concepts and cascading utilization pathways maximize the economic value and environmental benefits of pork fat feedstocks.

Waste-to-Energy Conversion: Waste streams, including wastewater, organic residues, and agricultural byproducts, can be co-processed with pork fat feedstocks in anaerobic digestion, gasification, or pyrolysis systems to produce biogas, syngas, or biochar for energy generation, heat recovery, and soil carbon sequestration. Combined heat and power (CHP) systems and anaerobic digesters offer synergies for co-producing biofuels and renewable energy from organic waste streams.

Closed-Loop Supply Chains: Closed-loop supply chain models integrate upstream feedstock production, biofuel processing, and downstream product distribution to minimize material losses, optimize resource utilization, and reduce environmental impacts. By establishing traceability, transparency, and accountability throughout the supply chain, closed-loop systems enhance sustainability performance and stakeholder confidence in pork fat biofuel products.

10.12.2 Advanced Biorefinery Concepts

Advanced biorefinery concepts leverage multi-feedstock processing, integrated conversion technologies, and product diversification strategies to enhance process flexibility, resilience, and value creation (Lamers et al. 2016). In the context of pork fat biofuels, advanced biorefineries incorporate:

Multi-Feedstock Flexibility: Biorefineries capable of processing multiple feedstocks, including pork fat, vegetable oils, waste oils, algae, lignocellulosic biomass, and organic residues, offer flexibility to adapt to changing market conditions, feedstock availability, and regulatory requirements. Co-feeding strategies, feedstock blending, and feedstock switching enable biorefineries to optimize production economics and product portfolios based on feedstock prices, availability, and sustainability criteria.

Integration of Conversion Pathways: Integration of multiple conversion pathways, such as transesterification, hydrogenation, pyrolysis, gasification, and microbial conversion, within a single biorefinery complex enables efficient utilization of feedstock resources, process synergies, and product diversification. Hybrid biorefinery configurations, process integration platforms, and modular reactor systems facilitate technology integration and scalability, allowing for tailored solutions to specific market demands and end-user requirements.

Co-Production of Value-Added Products: Co-production of value-added products, such as specialty chemicals, bio-based materials, nutraceuticals, and pharmaceuticals, alongside biofuels enhances revenue streams, market resilience, and value creation in biorefinery operations. Process integration, fractionation technologies, and product recovery strategies enable selective extraction of high-value compounds from pork fat feedstocks, expanding the range of marketable products and enhancing overall biorefinery profitability.

10.12.3 Digitalization and Industry

Digitalization technologies offer opportunities to optimize process efficiency, enhance productivity, and improve decision-making in pork fat biofuel production. Advanced data analytics, automation, and connectivity solutions enable:

Real-Time Process Monitoring and Control: Sensors, actuators, and IoT (Internet of Things) devices provide real-time data on process variables, equipment performance, and product quality parameters, allowing for proactive monitoring, predictive maintenance, and adaptive control strategies. Data-driven insights, machine learning algorithms, and artificial intelligence (AI) models optimize process conditions, mitigate risks, and improve overall operational efficiency in pork fat biofuel production facilities.

Supply Chain Optimization and Logistics: Digital supply chain platforms, blockchain technologies, and smart logistics solutions optimize feedstock procurement, transportation, and inventory management, reducing costs, delays, and supply chain disruptions. Supply chain visibility, transparency, and traceability enable stakeholders to track feedstock origins, production milestones, and product certifications, ensuring compliance with regulatory requirements and customer preferences.

Virtual Reality (VR) and Augmented Reality (AR) Training: VR and AR training simulations provide immersive learning experiences for operators, technicians, and maintenance personnel, facilitating skills development, troubleshooting, and safety training in pork fat biofuel production facilities. Virtual replicas of equipment, process units, and operational scenarios enable hands-on training, knowledge transfer, and continuous improvement initiatives, enhancing workforce competence and productivity.

Digital Twin Modeling and Simulation: Digital twin models of biorefinery processes, equipment, and assets enable virtual prototyping, scenario analysis, and performance optimization prior to physical implementation, reducing time-to-market, capital expenditures, and operational risks. Dynamic simulation tools, process optimization algorithms, and scenario planning software support decision-making, risk management, and strategic investment planning in pork fat biofuel production projects.

10.13 Market Diversification and Product Innovation

Market diversification and product innovation strategies expand the range of applications, end-users, and revenue streams for pork fat biofuels, enhancing market resilience and competitiveness (Ponnampalam and Holman 2023). Emerging trends in market diversification and product innovation include:

Sustainable Aviation Fuels (SAF): Growing demand for sustainable aviation fuels (SAF) presents opportunities for pork fat biofuel producers to enter the aviation market, which requires low-carbon, drop-in replacements for conventional jet fuels. Certification schemes such as the ASTM D7566 and the International Air Transport Association (IATA) SAF certification provide frameworks for verifying the sustainability and eligibility of bio-based jet fuels derived from pork fat and other renewable feedstocks.

Renewable Chemicals and Bioproducts: The transition towards a bio-based economy drives demand for renewable chemicals, bioplastics, and bio-based materials derived from pork fat and other biomass feedstocks. Applications include biodegradable plastics, lubricants, surfactants, solvents, and polymers, which offer environmental advantages, performance benefits, and market differentiation compared to petroleum-based counterparts.

Renewable Hydrogen Production: The emergence of renewable hydrogen as a clean energy carrier and energy storage solution creates opportunities for pork fat biofuel producers to diversify into hydrogen production and utilization. Electrolysis, steam reforming, and biomass gasification technologies enable the production of renewable hydrogen from water, steam, and biomass feedstocks, offering synergies with pork fat hydrogenation processes and decarbonization pathways for various sectors, including transportation, industry, and power generation.

Biogenic Carbon Offsets: The development of biogenic carbon offset markets provides opportunities for pork fat biofuel producers to monetize carbon sequestration and emissions reductions achieved through biofuel production processes. Carbon offset programs, carbon credit trading platforms, and voluntary carbon markets enable biofuel producers to generate additional revenue streams, attract investment, and demonstrate environmental stewardship through verified emissions reductions and carbon removals associated with pork fat biofuel projects.

10.14 Recommendations for Future Research

To advance the field of biofuel production from pork fat, further research is needed to address key challenges and explore new opportunities. The following recommendations outline potential areas for future investigation:

10.14.1 Feedstock Characterization and Optimization

Conduct comprehensive characterization studies of pork fat feedstocks to understand variations in composition, quality, and suitability for biofuel production. Investigate pretreatment methods to improve the quality and processability of pork fat feedstocks, including degumming, acidulation, and filtration techniques. Explore strategies for optimizing feedstock blending, co-processing, and co-feeding with other lipid sources to enhance feedstock flexibility, utilization rates, and cost-effectiveness.

10.14.2 Process Intensification and Integration

Develop novel reactor designs, process configurations, and integrated biorefinery concepts to improve process efficiency, energy recovery, and resource utilization. Explore advanced separation and purification techniques, such as membrane technologies, adsorption methods, and extraction processes, for enhancing product recovery and quality. Investigate opportunities for waste heat recovery, cogeneration, and integration with renewable energy systems to improve overall process sustainability and economics.

10.14.3 Catalyst Development and Catalytic Processes

Advance catalyst synthesis techniques, catalyst characterization methods, and catalyst screening approaches to identify new catalyst materials with enhanced activity, selectivity, and stability. Investigate heterogeneous catalytic processes, such as metal-supported catalysts, bifunctional catalysts, and shape-selective catalysts, for improving conversion efficiency and product yields. Explore catalytic upgrading strategies for converting intermediate bio-oil fractions, glycerol byproducts, and low-value co-products into higher-value fuels and chemicals.

10.14.4 Techno-Economic Analysis and Market Assessment

Conduct comprehensive techno-economic assessments (TEAs) and life cycle assessments (LCAs) to evaluate the economic viability, environmental sustainability, and social impacts of pork fat biofuel production technologies. Analyze market dynamics, feedstock availability, regulatory frameworks, and policy incentives to assess market opportunities and risks for pork fat biofuels in different regions and sectors. Investigate market diversification strategies, product differentiation opportunities, and value chain innovations to enhance market competitiveness and resilience.

10.15 Sustainability and Environmental Impact

Quantify the environmental footprint, carbon intensity, and land use implications of pork fat biofuel production pathways through rigorous life cycle assessments (LCAs) and environmental impact assessments (EIAs).

Assess the potential environmental co-benefits, ecosystem services, and social welfare impacts associated with pork fat biofuel production, including impacts on air quality, water resources, and biodiversity.

Explore strategies for enhancing sustainability performance, minimizing environmental risks, and maximizing social benefits through stakeholder engagement, community outreach, and participatory decision-making processes.

10.15.1 Policy Analysis and Regulatory Frameworks

Evaluate existing policy frameworks, regulatory incentives, and market mechanisms for promoting biofuel production from pork fat and other renewable feedstocks. Identify policy barriers, regulatory challenges, and market distortions that hinder the commercialization and adoption of pork fat biofuels, and recommend policy reforms and interventions to address them. Engage policymakers, industry stakeholders, and civil society organizations in dialogue and advocacy efforts to raise awareness, build consensus, and mobilize support for pork fat biofuel development and deployment.

10.16 Recommendations for Action

To translate the potential of pork fat biofuel production into tangible benefits for society, several actionable recommendations are proposed:

10.16.1 Investment and Funding

Governments, research institutions, and private investors should prioritize funding and investment in pork fat biofuel research, development, and demonstration projects. Funding initiatives should support interdisciplinary research collaborations, technology innovation, and pilot-scale testing to accelerate the commercialization of pork fat biofuel technologies.

10.16.2 Policy Support and Incentives

Policymakers should implement supportive regulatory frameworks, renewable fuel standards, and fiscal incentives to promote the production, distribution, and use of pork fat biofuels. Policy measures should include tax credits, production incentives, blending mandates, and public procurement programs to stimulate market demand and investment in pork fat biofuel projects.

10.16.3 Market Development and Infrastructure

Industry stakeholders should invest in infrastructure upgrades, retrofitting existing facilities, and expanding distribution networks to accommodate pork fat biofuel production and distribution.

Market development initiatives should focus on raising awareness, educating consumers, and fostering public acceptance of pork fat biofuels as a sustainable and viable energy source.

10.16.4 Research and Innovation

Research institutions, universities, and industry consortia should collaborate on research initiatives to advance feedstock characterization, process optimization, catalyst development, and sustainability assessment for pork fat biofuel production. Innovation hubs, technology incubators, and accelerators should support entrepreneurship, technology transfer, and commercialization efforts in the pork fat biofuel sector.

10.16.5 Capacity Building and Workforce Development

Educational institutions should offer specialized training programs, courses, and workshops on pork fat biofuel production, covering topics such as feedstock management, process engineering, and sustainability analysis. Workforce development initiatives should prioritize skills development, knowledge transfer, and career pathways in the bioenergy sector to cultivate a skilled workforce capable of driving innovation and industry growth.

10.17 Conclusion

The production of biofuels from pork fat represents a promising pathway towards sustainable energy production, waste valorization, and environmental stewardship. Through advancements in technology, policy support, market development, and collaboration, pork fat biofuels can play a significant role in mitigating climate change, enhancing energy security, and promoting economic development. This comprehensive exploration has highlighted the technical feasibility, economic viability, and environmental benefits of pork fat biofuel production. From feedstock characterization to process optimization, from policy incentives to market strategies, stakeholders across the bioenergy value chain have opportunities to contribute to the growth and success of this emerging sector. As we look to the future, it is essential to prioritize investment in research and development, policy support, market development, and collaboration initiatives to unlock the full potential of pork fat biofuels. By seizing these opportunities and addressing challenges through collective action and innovation, we can accelerate the transition to a more sustainable and resilient energy future. In closing, the journey towards sustainable bioenergy requires dedication, collaboration, and vision from all stakeholders. By harnessing the potential of pork fat biofuels and other renewable resources, we can pave the way for a cleaner, greener, and more prosperous world for generations to come. Together, let us embrace the challenge and opportunity of pork fat biofuels and work towards a brighter future for our planet and its inhabitants.

References

Amal R, Usman M (2024) A review of breakthroughs in biodiesel production with transition and non-transition metal-doped CaO nano-catalysts. Biomass Bioenergy 184:107158

Barla RJ, Anand A, Raghuvanshi S, Gupta S (2024) Life cycle assessment of renewable diesel production. In: Renewable diesel. Elsevier, Amsterdam, pp 65–86

Camilo GL, Queiroz A, Ribeiro AE, Gomes MCS, Brito P (2024) Review of biodiesel production using various feedstocks and its purification through several methodologies, with a specific emphasis on dry washing. J Ind Eng Chem 136:1

Dhir B (2024) Biofuel production from agricultural waste: a global trend. In: Emerging trends and techniques in biofuel production from agricultural waste. Springer Nature Singapore, Singapore, pp 1–13

Drewnowski A (2024) Perspective: the place of pork meat in sustainable healthy diets. Adv Nutr 18:100213

e Melo VM, Ferreira GF, Fregolente LV (2024) Sustainable catalysts for biodiesel production: the potential of CaO supported on sugarcane bagasse biochar. Renew Sust Energ Rev 189:114042

Ezeorba TPC, Okeke ES, Mayel MH, Nwuche CO, Ezike TC (2024) Recent advances in biotechnological valorization of agro-food wastes (AFW): optimizing integrated approaches for sustainable biorefinery and circular bioeconomy. Bioresour Technol Rep 26:101823

Ford JS, Bale CS, Taylor PG (2024) The factors determining uptake of energy crop cultivation and woodland creation in England: insights from farmers and landowners. Biomass Bioenergy 180:107021

Gbadeyan OJ, Muthivhi J, Linganiso LZ, Mpongwana N, Dziike F, Deenadayalu N (2024) Recent improvements to ensure sustainability of biodiesel production. Biofuels:1–15

Gonçalves PC, Monteiro LPC, de Sousa Santos L (2020) Multi-objective optimization of a biodiesel production process using process simulation. J Clean Prod 270:122322

Ismaeel HK, Albayati TM, Al-Sudani FT, Salih IK, Dhahad HA, Saady NMC et al (2024) The role of catalysts in biodiesel production as green energy applications: a review of developments and prospects. Chem Eng Res Des 204:636

Iyke BN (2024) Climate change, energy security risk, and clean energy investment. Energy Econ 129:107225

Jaiswal KK, Chowdhury CR, Dutta S, Banerjee I, Jaiswal KS, Nisansala HMD et al (2024) Synthesis of renewable diesel as a substitute for fossil fuels. In: Renewable diesel. Elsevier, Amsterdam, pp 1–31

Jambo SA, Abdulla R, Azhar SHM, Marbawi H, Gansau JA, Ravindra P (2016) A review on third generation bioethanol feedstock. Renew Sust Energ Rev 65:756–769

Kesarwani S, Kumar M, Tripathy DB, Gupta A, Kumar S (2024) Oil and fats as raw materials for industry: an introduction. In: Oils and fats as raw materials for industry. Wiley, New York, pp 1–32

Lamers P, Searcy E, Hess JR (2016) Transition strategies: resource mobilization through merchandisable feedstock intermediates. In: Developing the global bioeconomy. Academic Press, Amsterdam, pp 165–185

Lin CY, Lu C (2021) Development perspectives of promising lignocellulose feedstocks for production of advanced generation biofuels: a review. Renew Sust Energ Rev 136:110445

Lugani Y, Brar SK, Kaur Y, Singh BP, Kumar D, Kumar S (2024) Sustainable production of advanced biofuel and platform chemicals from woody biomass. In: Sustainable biorefining of woody biomass to biofuels and biochemicals. Woodhead Publishing, Sawston, pp 163–194

Morone P, Cottoni L, Giudice F (2023) Biofuels: technology, economics, and policy issues. In: Handbook of biofuels production. Woodhead Publishing, Sawston, pp 55–92

Nwokolo NL, Enebe MC (2024) An insight on the contributions of microbial communities and process parameters in enhancing biogas production. Biomass Convers Biorefinery 14(2):1549–1565

Ponnampalam EN, Holman BW (2023) Sustainability II: sustainable animal production and meat processing. In: Lawrie's meat science. Woodhead Publishing, Sawston, pp 727–798

Praveena V, Martin LJ, Matijošius J, Aloui F, Pugazhendhi A, Varuvel EG (2024) A systematic review on biofuel production and utilization from algae and waste feedstocks—a circular economy approach. Renew Sust Energ Rev 192:114178

Rabbani M, Hosseini A, Karim MA, Fahimi A, Karimi K, Vahidi E (2024) Environmental impact assessment of a novel third-generation biorefinery approach for astaxanthin and biofuel production. Sci Total Environ 912:168733

Ramalingam G, Priya AK, Gnanasekaran L, Rajendran S, Hoang TK (2024) Biomass and waste derived silica, activated carbon and ammonia-based materials for energy-related applications—a review. Fuel 355:129490

Saidu MM, Lawal EE, Tsado PY, Yakubu JG, Adeniyi OS, Oyewole OA, Dabai AI (2024) Potentials of organic waste to provide bioenergy. In: Microbial biotechnology for bioenergy. Elsevier, Amsterdam, pp 179–218

Shahzad K, Cheema II (2024) Low-carbon technologies in automotive industry and decarbonizing transport. J Power Sources 591:233888

Sharma Y, Shankar V (2024) A comparative assessment of microbial biodiesel and its life cycle analysis. Folia Microbiol:1–27

Shweta, Capareda SC, Kamboj BR, Malik K, Singh K, Bhisnoi DK, Arya S (2024) Biomass resources and biofuel technologies: a focus on Indian development. Energies 17(2):382

Singh N, Saluja RK, Rao HJ, Kaushal R, Gahlot NK, Suyambulingam I et al (2024) Progress and facts on biodiesel generations, production methods, influencing factors, and reactors: a comprehensive review from 2000 to 2023. Energy Convers Manag 302:118157

Skaggs RL, Coleman AM, Seiple TE, Milbrandt AR (2018) Waste-to-energy biofuel production potential for selected feedstocks in the conterminous United States. Renew Sust Energ Rev 82:2640–2651

Sun Y (2024) Technology research and development prospects of biofuels. J Educ Educ Res 7(1):11–15

Sungur Ş (2024) Biofuels. In: Handbook of emerging materials for sustainable energy. Elsevier, Amsterdam, pp 399–417

Vongsawasdi P, Noomhorm A (2014) Bioactive compounds in meat and their functions. In: Functional foods and dietary supplements: processing effects and health benefits. Wiley, New York, pp 113–138

Wang J, Azam W (2024) Natural resource scarcity, fossil fuel energy consumption, and total greenhouse gas emissions in top emitting countries. Geosci Front 15(2):101757

Yu Y, Li Q, Bao Y, Fu E, Chen Y, Ni T (2024) Research on the measurement and influencing factors of carbon emissions in the swine industry from the perspective of the industry chain. Sustainability 16(5):2199

Pig Farming and Business Opportunities for Financial Benefit

11

Saroj K. Rajak, Jaya Bharati, Satish Kumar, Rakhi Bharti, and Pinky Preety

Abstract

Pig farming is a significant component of the livestock sector and there exists demand-supply gap for pork in the country. The landscape of pig production in India is dominated by small-scale and backyard system of pig rearing. Looking into the burgeoning population, urbanization pattern and change in food habits including the quick returns from pig farming, piggery has great potential to support livelihood and nutritional security. In the recent decades, piggery sector has gained momentum and is becoming increasingly popular among agripreneurs as a viable business opportunity for financial benefit. There exists limited number of scientifically managed commercial farms, slaughterhouse, pork processing facilities, pig value chain operators and piggery-associated enterprise in India, which offers great prospect for economic turnaround. Nevertheless, issues like occurrence of transboundary emerging and re-emerging diseases of pig, which causes high mortality need to be addressed. The intensification of pig rearing system would bring in waste management and environmental issues, which need to be integrated and addressed with the piggery development schemes. Looking into the challenges and untapped potential of piggery sector in India, pig rearing

S. K. Rajak (✉)
Department of Veterinary and A.H. Extension Education, Bihar Veterinary College, Patna, Bihar, India

J. Bharati · S. Kumar
ICAR—National Research Centre on Pig, Guwahati, Assam, India

R. Bharti
Department of Veterinary and A.H. Extension Education, CoVAS, Kishanganj, Bihar, India

P. Preety
Department of Veterinary and A.H. Extension Education, GADVASU, Ludhiana, Punjab, India

© The Author(s), under exclusive license to Springer Nature Singapore Pte Ltd. 2024
T. Rana, B. Soto-Blanco (eds.), *Good Practices and Principles in Pig Farming*, Livestock Diseases and Management,
https://doi.org/10.1007/978-981-97-4665-1_11

can be developed as an enterprise by possible interventions at different level of pig production chain and supporting farmers on technical, financial and marketing grounds.

Keywords

Piggery · Breeding · Management · Farrowing · Pork · Economics

11.1 Introduction

Piggery plays a crucial role in ensuring food, nutritional and livelihood security of socio-economically weaker people and tribal communities in particular. India is endowed with rich biodiversity of pig breeds and pig husbandry is a part of integrated farming system throughout the country. Pig farming is highly unorganized in India and majority of them is raised in small holder, low input production system (DAHD 2022). Piggery is more concentrated in eastern and north-eastern part of India in the states of Assam, Jharkhand, Meghalaya, West Bengal, Chhattisgarh, Sikkim, Tripura, Nagaland, Manipur, Arunachal Pradesh and Mizoram. These states have favourable agro-climatic conditions and cultural acceptance of pork consumption. Small-scale and backyard pig farming are quite common in these regions, which provides additional income to the farmers. As per the 20th Livestock census, the total pig population in the country is 9.06 million which accounts for 1.70% of the total livestock population and contributes to 3.85% of total meat production in the country (BAHS 2023). The contribution seems to be small, but looking into the pork demand-supply gap, the potential of this sector is yet to be explored in India. While India is a major exporter of various agricultural products, pork export from India is currently limited. The country has the potential to increase pork exports by improving infrastructure, implementing quality control measures and complying with international standards. The major domestic market for pork is north-east India. Yet in most part of the country, piggery has abundant options to be adopted like an independent venture, integrated fish-pig farming, pig feed, supplements and biologicals manufacturing, value education of pork products etc.

In the recent past, piggery sector in India has gained popularity as a profitable livestock venture mainly due to its economic viability and adaptability to low input farming system (Bharati et al. 2022). Pig rearing, which was predominantly a small-scale enterprise, practised in the form of backyard piggery by small and marginal farmers of a particular community (Sahu 2022), has now started attracting interest and investment from agripreneurs and industry. There has been a perceptible increase in establishment of commercial piggery farms in states like Punjab, West Bengal, Uttar Pradesh, Maharashtra and Rajasthan. Slowly and steadily, piggery sector farming is gaining momentum in the country due to numerous reasons. The market demand for pork and value-added processed pork products is growing in India, mainly due to the change in dietary likings, increasing urbanization and rising middle-class income (Bhaduria et al. 2023). The increasing presence of many international restaurant chains throughout the country further enhanced the popularity of

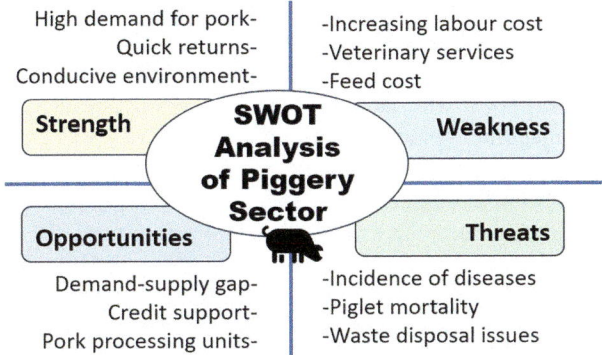

Fig. 11.1 The graphical representation of strength-weakness-opportunity-threat (SWOT) analysis of the piggery sector in India

pork-based specialties like bacon, sausages, salami, frankfurters, momos and other processed pork products. Thus, a conducive environment and market opportunity is existing, favouring the development of the commercial pig farms to supply to a large consumer base (NCDC 2017). Nevertheless, the credit, institutional and capacity building support from government and research institutions attracted youths and entrepreneurs which promoted the sector and attracted entrepreneurs. The government has been boosting the piggery sector as an avenue for secondary agriculture and integrated farming system among rural masses and unemployed youths. Hence, pig farming has transformed into the most sustainable livestock-based agri-business enterprise, which a farmer can adopt as per the feasibility and available resources. The SWOT analysis of the piggery sector in India is represented diagrammatically in Fig. 11.1.

11.2 Advantages of Pig Farming as an Enterprise

Pigs are hardy animals and different indigenous and cross-bred pigs are available in India that thrive well under different agro-climatic conditions throughout the country (Kadirvel et al. 2023). They can be reared almost anywhere with minimal housing and management requirement. Pigs can digest a broad range of food sources and hence can be reared even under zero-input system. Pigs are the most effective feed converters after broilers among the many livestock species, which means they have high feed conversion ratio making piggery as the most promising source for the production of meat. Pigs are the most prolific breeders with high fecundity rate (litter size of 7–14) and shorter generation interval (<2 years). Piggery offers quick returns since the market weight of 60–90 kg can be obtained in 7–8 months of age. In addition to nutritious pork, it also provides bio-manure for agricultural farm, bristles for brushes and animal fat for soap and chemical industry. There exists good demand for pig products in domestic market and there is huge potential for export to international markets. Seasonally employed farmers can engage in pig farming and increase their income, which can contribute greatly to raise their socio-economic status (Kadirvel et al. 2023). In India, pig farming is the primary source of income for the rural poor, who lack the resources to engage in scientific farming with improved management practices (Singh et al. 2023). However, the commercial pig sector generates revenue for farmers, entrepreneurs and rural communities. Employment opportunities exist across the value chain of pig production, from piglet production to pork processing and marketing. Furthermore, the commercial pig sector in India has seen technical advances in breeding, feeding and management methods. Because of the nation's enormous potential for improving pig production, central government and several state governments have launched several schemes for credit and technical support for pig farmers. A conducive environment for economic pig production has been developed in the country, which can make piggery a more profitable and sustainable enterprise by utilizing cutting-edge technology in conjunction with scientific management to boost productivity, feed efficiency and disease control.

11.3 Different Types of Pig Farming System

Three different types of pig farming practices are prevalent in India, namely, small-scale backyard, medium-scale and large-scale commercial system (Bora et al. 2020). Integrated pig farming is also popular in India, where pig is integrated with other components of agriculture. The adoption of these farming systems depends on the type of pig breed, local demand, credit support, facilities available, market and personal choice of the farmers.

11.3.1 Integrated Pig Farming

Integrated pig farming is prevalent throughout the country in various combinations viz. pig-fish system, pig-fish-duck system, pig-fish-vegetable crop system, pig-fish-fruit crop system, pig-biogas-fish system etc. wherein pig forms the livestock component of the integrated farms (Nhu et al. 2015). The pig-fish integrated farming is most popular in the country, where pig dung contributes to 70% digestible food for fish and thus reduces the 60% of input cost in fish culture. The pig dung increases the biological productivity of the pond and fishes can feed on pig waste and let over feed. This system economically utilizes the land availability, since the pigsties are built over the pond embankments. The pond can also be used to rear ducks for meat and eggs. Silt deposited in ponds can be utilized as a fertilizer for horticultural crops (Tripathi and Sharma 2001). The pig-fish integrated farming system is represented graphically in Fig. 11.2.

Fig. 11.2 The graphical representation of a pig-fish integrated farming system; wherein different components of integrated farming system have been indicated as (1) Pig, (2) Fish, (3) Duck, and (4) Horticultural crops

11.3.2 Small-Scale Backyard System

This is a traditional pig rearing system usually practised in rural and semi-urban areas. Under this system, two to ten pigs are kept in open areas, often in the backyard of the house and are free to scavenge. The animals are sometimes given a temporary shed during rains and fed on kitchen and crop waste. There is absence of scientific pig management practice. The animals exhibit low body weight gain, less litter size and high pre-weaning mortality. Backyard system is more prevalent in the eastern and north-eastern part of India, mainly due to the deep association of pork with the dietary habits of people in this region (Kadirvel et al. 2013). The indigenous pig breeds like Ghungroo, Mali, Doom etc. which can thrive well on kitchen waste and have a culinary preference due to their black colour are mainly reared under this system (Bharati et al. 2023a; Mohan et al. 2023). However, with the increasing awareness on scientific pig farming techniques, backyard piggery is steadily being replaced by semi-intensive farming system.

11.3.3 Medium-Scale Pig Production

In this type of production system 10–50 pigs are reared by adopting a semi-intensive rearing practice. Locally available cross-bred, indigenous or non-descript pigs are kept in sheds made up of cheap and easily available fencing materials. There is also a provision of an open area for free movement of animals. Pigs are fed on a mixture of commercially available balanced ration and locally available feed resources like banana stem, cassava, colocasia, tapioca, potatoes, cauliflower waste, rice gruel etc.

They are also fed with properly cooked kitchen, hotel, vegetable and fruit market waste etc. which sometimes serves as a major feed for pigs. Pig farmers may apply healthcare measures in pigs, like deworming, vaccination, iron injection; however, the scientific pig rearing is lacking in respect of breeding and housing management. This system is often practised as an all-in all-out system and is associated with finisher farms, wherein the farmers purchase piglets from live pig market and are kept for fattening. When the animals reach 50–70 kg weight they are usually sold off in the local market after one production cycle. The medium-scale pig farms can be easily transformed into a medium-scale commercial pig farm with 50 parent stock pigs and inclusion of scientific methods of pig farming.

11.3.4 Large-Scale Pig Production

This is the most advanced type of pig production system, housing 50–800 pigs raised on intensive system of management and scientific rearing methods are applied for each production stage. They are common in urban and peri-urban areas with large commercial prospects for pig farming. This system is specifically developed as breeding or a seed farm and finisher farm. The farm consists of expensive facilities and trained staff. Pigs are usually confined to their shed. Scientific approach is applied in this type of system for production of safe pork. The reproductive and productive performance of pigs are usually high, due to the involvement of standardized management procedures, scientific breeding, preventive healthcare and good quality feed (Maes et al. 2020). These farms may also include boar semen production and processing units for practicing artificial insemination in sows. These farms usually house highly prolific exotic or cross-bred pigs. These farms serve as supplier farm for piglets, breeding gilts/sows and boars to other small farms in the adjoining areas. They also supply finishers and culled animals to the slaughterhouses and to the pig traders who transport pigs to the distant geographical locations in the country where the demand of pork is high.

11.4 Establishment of a Pig Farm

A productive pig farm requires a substantial investment on founder stock of breeding animals, type of housing, healthcare and availability of feed. The type of farming system to be adopted depends on the availability of initial capital, land area, number of sow units, inclusion of slaughter, pork processing facility and size of feed storage unit. Following critical points need to be considered while planning for establishment of a commercial pig farm.

11.4.1 Selection of Breeds

The selection of pig breed along with effective breeding management is per-requisite for good economic return. The productive and reproductive performance of an animal is critically determined by the breed in conjugation with the management followed in the farm. The National Guidelines for formulation of State Pig Breeding Policy in India, laid down by the Department of Animal Husbandry and Dairying, Government of India, has provided the flexibility to the States to devise upon a pig breeding policy as per their regional requirement within the framework outlined in the guideline. It is formulated with the objective to improve the genetics of local pigs, conserve and maintain the indigenous germplasm in their domestic tract, maintain scientifically bred defined cross-bred pigs at farmer's field, expand and maintain the breeding infrastructure, propagate elite pig germplasm through technological interventions like artificial insemination. The guidelines also aim at holistic development of piggery sector through breeding, feeding, housing management, value addition and marketing. The guidelines on breeding pyramid has been developed, which consists of nucleus sire and dam lines at the top level, multiplication cross-bred females at the middle level and commercial piglet production and finishers at the bottom level (DAHD 2022).

The selection of pig breed for cross-breeding and propagation of cross-bred animals has been planned taking into account the 15 different agro-climatic zones as delineated by the Planning Commission of India which is based on physiography and climate delineating that region. A particular zone exhibits similar features of climatic, soil, physiography and cropping patterns leading to uniformity in practices pertaining to agricultural and allied sector economic activities (Rai et al. 2008). A highly prolific exotic breeds may not be adaptable under the given agro-climatic region and may not yield good litter size and carcass weight, even under scientifically managed commercial set-up. Hence, the cross-bred pigs have been developed in different states of the country, keeping in view the high prolificacy, growth rate, meat quality of exotic pigs and the adaptability of indigenous pigs, under All India Co-ordinated Research Projects on Pig (AICRP) with ICAR-National Research Centre on Pig, Guwahati, as nodal centre. Studies on these cross-bred pigs indicate that they perform well under the given agro-climatic condition and has been widely accepted by farmers. In a smallholder production system adoption of cross-bred pigs over the indigenous pigs can yield more economic benefits, as found in the study by Kadirvel and co-workers in the Eastern Himalayan hill region of India (Kadirvel et al. 2023). They studied the productive and reproductive traits of the Lumsniang cross-bred pig variety, developed by cross-breeding exotic Hampshire and indigenous Niang Megha pigs. The cross-bred pigs performed better and were more adaptable to the hill ecosystem, as compared to the prevailing local indigenous pigs in the region. The cross-bred pigs performed well even in the low-input backyard production system (Kadirvel et al. 2023). Thus, selection of breed which suits well to the type of production system and agro-climatic zone is a key factor in economic viability of the pig farm.

11.5 Type of Housing

The Bureau of Indian Standards has described the code of practice for pig housing and has specified dimensions of various structures of pigsty so as to cater to the scientific housing of pigs in the country. Accordingly, the pig shed should be minimum 15 m away from human dwellings and factories, 30 m away from dairies, food grain storage centres, poultry and other animal farms and 45 m away from sources of fire like industry furnaces etc. It should also be 1 km away from garbage dumping areas, slaughterhouse of other animals, hide curing centres and tanneries (IS 3916:1966). Pig housing should be fenced from all sides and should provide protection from extremes of weather condition, predators and water logging. It should be well connected with the roads; however, it should be located at least 50 m away from the nearest transit roads. The long axis of the shed should be built in the direction, so that plenty of sunlight and ventilation is available. The floor and walls should be sturdy enough so as to withstand rooting by pigs. The BIS recommends the provision of both covered floor area and open yard area adjacent to its sty for each animal. The space requirements of different categories of pigs are mentioned in Table 11.1. The animals should be kept in sties, which can be divided into one or more pens, depending upon the number of pigs housed either in groups or individually, which in turn is determined by age, sex and purpose of rearing. The number and dimensions of pens in a sty depend upon the number of animals reared and floor area requirement of the category of pig. This provision is quite economic for a small-scale production system, in which different categories of animals can be housed under the same roof. The wallowing tank can be constructed in pig farms located in hot agro-climatic regions. It can be made of concrete cement. Instead, overhead sprinklers or foggers can be installed in the sheds to ward of heat stress in pigs. The schematic diagram representing the set-up of a pig farm with dimensions of different categories of pen viz. boar pen, dry sow pen, weaner, fattening pen is shown in Fig. 11.3.

In a large-scale commercial production system, which houses more than 100 pigs, provision of different types of sties viz. boar sty, dry sow and gilt sty, farrowing or nursing sow sty, fattening sty, weaner sty and sick sty for housing different

Table 11.1 The housing dimensions of covered floor area and open yard area per animal for different types of animals in a pig farm

S. no.	Type of animal	Covered floor area/ animal (m^2)	Dimensions (m)	Open yard area/ animal (m^2)
1.	Boar	6.25–7.50	2.5 × 2.5 to 2.5 × 3.0	8.8–12.0
2.	Farrowing sow	7.50–9.00	2.5 × 3.0 to 3.0 × 3.0	8.8–12.0
3.	Weaner/fattening pig	0.96–1.80	0.8 × 1.2 to 1.2 × 1.5	8.8–12.0
4.	Dry sow/gilt	1.80–2.70	1.2 × 1.5 to 1.8 × 1.5	1.4–1.8

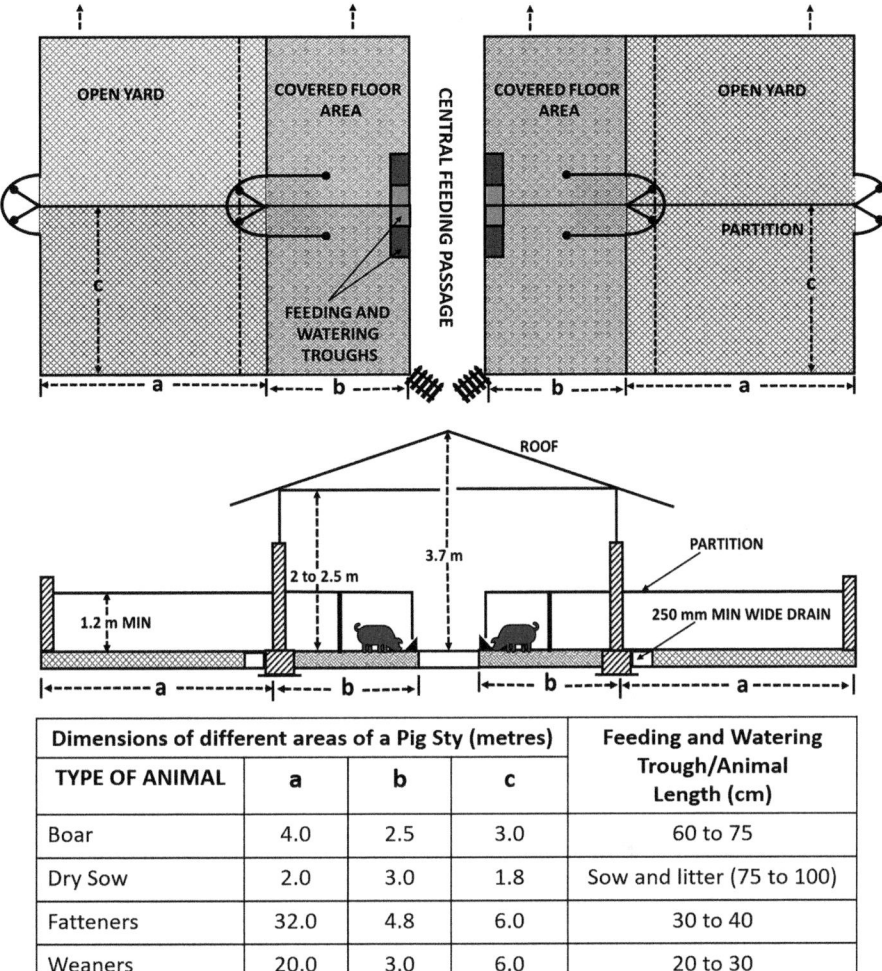

Fig. 11.3 The schematic diagram representing the setup of a pig farm with the dimensions of different categories of pen, viz., boar pen, dry sow pen, weaner, and fattening pen. Reference: Code of practice for pig housing by The Bureau of Indian Standards (IS 3916:1966)

categories of pigs can be made for better management and reducing labour cost. Commercial piggery unit shall also include weighing yard, store building, feed store, holding pen, loading and unloading ramp, manure pit, garbage boiling vat and boiler room or any other waste disposal system. The schematic diagram representing a large-scale commercial set-up with provision for different types of sty is shown in Fig. 11.4.

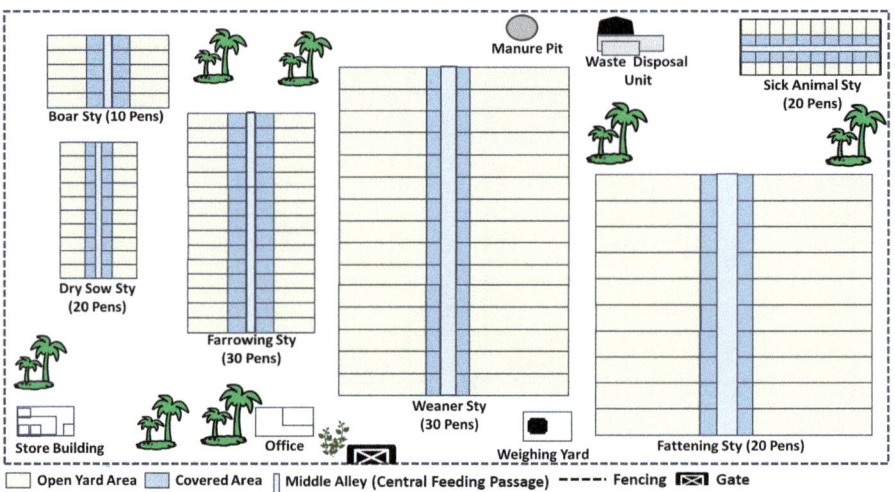

Fig. 11.4 The schematic diagram of a large-scale commercial pig farm setup with provision for different types of sties and supportive units. Reference: Code of practice for pig housing by The Bureau of Indian Standards (IS 3916:1966)

11.6 Reproductive Management

Regular screening of sows and gilts should be conducted for determining the day of oestrus and insemination. The gilts should be bred only when they reach 70 kg weight. Flushing, that is providing extra feed to gilts (0.5–1 kg per day) 10–15 days prior to mating in order to provide more energy should be done, to increase the number of ovulations and thus the litter size in pigs. Farmers should be well trained in determining the symptoms of heat and standing reflex in animals. Feeding during pregnancy in pigs should be regulated, so that the animals are not obese or thin. Pregnant animals should be housed separately for at least 2 weeks before the likely date of farrowing. The farrowing shed should be disinfected appropriately before transferring the pregnant animal. There should be provision of light bedding material and sufficient clean water in the shed. The sow/gilt should be constantly monitored during the last few days and veterinarian should be consulted on case of difficulty in farrowing. The placenta should be immediately disposed and care should be taken that it is not eaten by the sow. These parameters are critical for a profitable pig farming enterprise, since the reproductive parameters of the gilts/sows determines the economic returns (Bharati et al. 2023b). The gilts/sows can be inseminated either by natural mating or artificial insemination. In either of the case, indiscriminate breeding of pigs should be avoided, since it may lead to inbreeding, which ultimately leads to decrease in litter size, high pre-weaning mortality causing economic losses in a pig farm.

11.7 Health Management

Proper health calendar should be maintained in the farm. The piglets should be administered iron injection at 4th and 14th days of age to avoid piglet anaemia. The piglets should be dewormed at weaning and then repeated at 3–4 months intervals. Timely vaccinations for classical swine fever and foot and mouth disease should be administered to pigs as a preventive measure to avoid mortality due to diseases (Kumar et al. 2023). Proper biosecurity should be maintained in pig farm so as to avoid deadly diseases caused by African swine fever virus, porcine circovirus, porcine respiratory and reproductive syndrome virus etc. (Deb et al. 2022). Periodical tests for *Brucellosis*, *Leptospirosis* and other notified diseases should be conducted in breeding boars.

11.8 Maintaining Hygiene and Biosecurity

Pigs of all age groups should be kept under sanitary conditions to prevent occurrence of diseases and development of foul smell. Biosecurity, which is the application of defined management practices that decrease the chances for infectious agents to enter or spread in an animal farm. Adoption of both external and internal biosecurity measures in pig farm is important to restrict pathogens from gaining access to the herd and also preventing the spread of pathogens from an infected to healthy animals within the herd. Adoption of biosecurity measures helps in improving the productive efficiency of the farm, by reducing the incidence of diseases, and thus reducing the expenditure on veterinary aid and loss of animals due to high mortality. It may contribute significantly in decreasing the use of antibiotics and thus prevent the occurrence of anti-microbial resistance (Alarcón et al. 2021).

11.8.1 Record Keeping

Record keeping of farm activities is most important for evaluation as well as for the improvement of the production and reproduction in pigs. Recording of different parameters pertaining to production, reproduction, treatment, vaccination, feed, culling, labour etc. is very essential to track the functionalities and economic efficiency of a piggery unit. There should be a separate livestock register with details of date of birth, age at purchase, sex, source of animals, litter size, weaning age and weight, disease, vaccination, deworming etc. An individual health card and quarantine record can also be maintained. Recording of information on ancestry, i.e. pedigree of pigs is most critical information for devising breeding plan in a farm. A register for feed and fodder stock and purchase should be maintained. Records for mortality, reason for culling and post-mortem findings should be maintained. There should be register maintained at the farm gate for records of vehicle and human movements along with the technical and non-technical staff and labour record should also be maintained in a large-scale commercial pig unit.

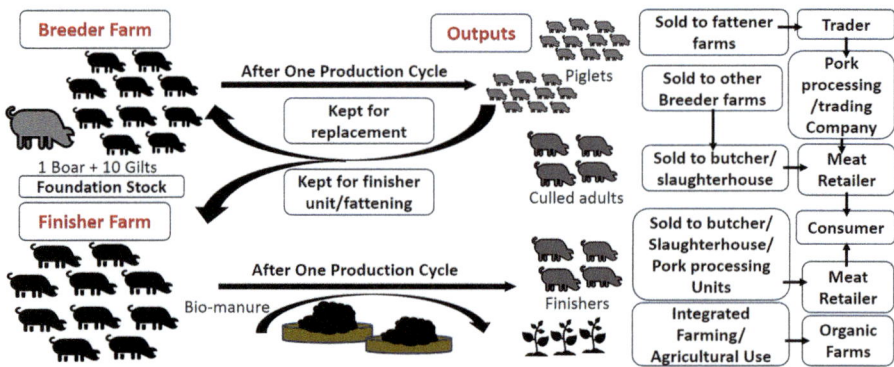

Fig. 11.5 The components of breeder and finisher farms, their output and different market components along the pig value chain during one production cycle

11.9 Economics of Pig Farming

The annual income or economics of pig farming depends on many factors like model of farming viz. backyard system or commercial farming, number of pigs reared, breed of pig, feeding practices and market options. The income also varies from farmers who are exclusively involved in breeding farms or fattening farms to farmers who are engaged in mixed type of farms which maintains both the type of farms concurrently. Moreover, in small-scale and medium-scale farming system, the inclusion or exclusion of family labour and kitchen/crop waste as feed gives significant difference to the annual return from pig farming. The utilization of pig waste for generating biofertilizer can add to the income and benefits for agricultural purpose. The income mostly varies on the point of pig value chain, wherein the farmers sell his output, namely, butchers, middle-men, pork traders, other breeder farms or slaughterhouses. The different components of pig production, output and market avenues during one production cycle is depicted graphically in Fig. 11.5.

The different parameters taken for calculating the annual income from pig farming is shown in Table 11.2. The economics of pig farming for 20 sow unit (20 sows +2 boar) for breeding farm is detailed in Table 11.3.

11.10 Policy and Credit Support

Multiple schemes and policies have been launched by the Central and State governments to support pig farmers and entrepreneurs for establishing modern commercial pig farms. One such scheme, National Livestock Mission (NLM) of Department of Animal Husbandry & Dairying, Government of India, has been implemented to meet the present need of the animal husbandry sector which aims towards employment generation, entrepreneurship development and increase in per animal productivity. The NLM Scheme also aims for entrepreneurship development so as to create

Table 11.2 The parameters taken for calculating the annual income from pig farming

Sl. no.	Parameters	Value
1.	No. of sows/gilt reared (no.)	20
2.	No. of boars maintained (no.)	2
4.	Cost of Sow or Gilt (in Rs.)	7000
5.	Cost of Boars (in Rs.)	8000
6.	Selling price of culled sow (in Rs.)	11,000
7.	Sale price of culled boar (in Rs.)	10,000
8.	No. of farrowings per year (no.)	2
9.	Litter size of sow (no.)	10
10.	Mortality among piglets (weaners)	5%
11.	Mortality among fatteners	2%
12.	Weaning period (days)	45
13.	Gestation period (days)	114
14.	Feed requirements per piglets	1 kg per day
15.	Feed requirements per boar/sow	2 kg per day
	Feed requirements per lactating sow	2.5 kg per day
	Feed requirements per dry sow	1.5 kg per day
	Total creep feed per piglet during suckling	2 kg
16.	Cost of feed	30/kg
	Cost of creep feed	35/kg
Space requirement (m²/animals)		
17.	Boar	21
18.	Lactating sow with its piglets	21
19.	Dry sow	3.6
20.	Fattener of 3–5 months age	1.5
21.	Fattener of 6–8 months age	1.8
22.	Store room (m²)	19
23.	No. of gunny bags per ton of feed	20
24.	Income from gunny bags (Rs./bag)	10
25.	No. of gilt/sow housed together	5
26.	No. of boar housed per pen	1
27.	Age at sale of piglets	After weaning

the forward and backward linkage for the animal produce available in the unorganized sector and to link them with the organized sector. The mission will be implemented through the State Implementing Agency established under the State Animal Husbandry Department, which in turn should establish their State Implementing Agencies or identify the agency already established for implementation of the NLM. The subsidy amount will be directed through the Small Industries Development Bank of India (SIDBI). Under NLM, there is a dedicated activity on "Promotion of Piggery Entrepreneur", with the aim for creation of entrepreneurship through one-time capital subsidy to individuals or self-help groups (SHG), Framers Producer Organizations (FPO)/Farmers Cooperatives (FCOs)/Joint Liability Groups (JLGs) and Section 8 companies. The entrepreneur will be provided assistance for the establishment of a breeder farm with minimum 100 sow and 25 boars from the

Table 11.3 The economics of a 20-sow unit (20 Sow + 2 Boar) crossbred commercial pig farm

Sl. no.	Particulars	Total area (m²)	Total cost (in Rs.)
Non-recurring cost			
1.	Cost of foundation stock		156,000
2.	Sty for gilt/sows (1.8 m² covered area and 1.8 m² open area) (4 pens)	72	
3.	Sty for boar (9 m² covered area and 12 m² open area)	42	
4.	Farrowing sty (9 m² covered area and 12 m² open area) (8 pens)	168	
5.	Weaner/grower sty (0.9 m² covered area and 0.9 m² open area) (16 pens)	288	
6.	Sick animal sty	10	
7.	Total floor space (2 + 3 + 4 + 5 + 6) @500/m²	586	293,000
8.	Cost of equipment (Rs. 500/adult)		11,000
	Total non-recurring expenditure		460,000
	Loan required (90% of non-recurring expenditure)		368,000
	Self-investment (10% of non-recurring expenditure)		92,000
	Repayment in five equal instalments		73,600
Recurring cost			

Sl. no.	Particulars	Cost (in Rs.)
A. Fixed cost		
1.	Interest on loan @12% per annum	44,160
2.	Depreciation on animals (5%)	7800
3.	Depreciation on building (10%)	29,300
4.	Depreciation on equipment (10%)	1100
5.	Insurance of animals (7%)	10,920
	Total Fixed cost	**93,280**
B. Variable cost		
1.	Feeding cost	
a.	20 gilt and 2 boars till maturity (120 days)	79,200
b.	2 breeding boars for 230 days	27,600
c.	Gilts during gestation	136,800
d.	Lactating sows	67,500
e.	Creep feed for 200 piglets	14,000
f.	Grower feed for 190 weaners up to 7 months (165 days)	940,500
g.	Sows after weaning of litters till pregnancy (30 days)	27,000
h.	Gilts during gestation	136,800
i.	Lactating sows	67,500
j.	Creep feed for 200 piglets (second batch)	14,000
2.	Labour cost (6000 per month)	72,000
3.	Veterinary cost (vaccine and medicine)	10,000
4.	Miscellaneous cost	5000
	Total variable cost	**1,597,900**

(continued)

Table 11.3 (continued)

Total recurring cost (A + B)	1,691,180
C. Income from all resources	

Sl. no.	Particulars	Cost in Rs.
1.	Sale of 190 growers having 80 kg bwt @150 per kg	2,280,000
2.	Sale of 200 weaners @5000 per piglets	1,000,000
3.	Sale of manure and gunny bags	20,000
Total returns		3,300,000
Gross income (C − Total recurring)		1,608,820
Net Income (Gross income − Loan repayment)		1,535,220
Net income per month (Rs./month)		85,290

Central or State Government/university farms or local farmers with high genetic merit. The scheme will provide 50% up to Rs. 30 lakh capital subsidy towards the capital cost of the project, which includes expenditure on housing, breeding, equipment/machines, purchasing animals along with transportation and insurance cost. No subsidy will be provided for purchase/lease of land (https://nlm.udyamimitra.in/Default/ViewFile/?id=NLMOperationalGuidelines.pdf&path=MiscFiles). The Animal Husbandry Infrastructure Development Fund (AHIDF) under Atma Nirbhar Bharat Abhiyan stimulus package had been initiated by the Government of India for incentivizing investments by individual entrepreneurs, private companies, micro-small and medium enterprises, Farmer Producer Organizations (FPOs) and Section 8 companies in the field of animal husbandry in India. It aims to provide support for establishing pig breeding and fattening farm with modern infrastructure, integrated production system, pig semen station and artificial insemination technology centres including pork processing and value addition infrastructure (AHIDF 2020). The pig breeding policies have been developed by major pig rearing states of the country like Assam, Mizoram, Meghalaya, Arunachal Pradesh, Nagaland, Himachal Pradesh, with due consideration of their social and cultural system. These policies establish guidelines for scientific breeding and conservation of elite germplasm, streamlining bank credit facility, subsidy for small holder pig farmers, subsidy on adoption of artificial insemination and waste management, pork processing, infrastructure development and tax holidays for commercial pig farm units. The Government of Odisha, Fisheries and ARD Department, has laid down the guidelines of the State Plan Scheme, Integrated Livestock Development Programme, Small Animal Development Centre for the establishment of 10 + 1 piggery unit (10 sow +1 boar) for semi-commercial pig farming to support small entrepreneurs for income generation. Under this scheme the individual was provided with 40% subsidy and rest was to be arranged either through bank loan or self-finance (https://fard.odisha.gov.in/sites/default/files/2023-03/13.%20Semi%20Commercial%20Pig%20Farming%20Guideline%202021-22.PDF).

11.11 Possible Interventions in the Piggery Sector

Introduction of region-specific recommended breed (cross-breed of indigenous) nuclear breeding farms at each district level in cooperative mode or in collaborative mode with Animal Husbandry Departments can cater to the need of pig farmers with respect to availability of good quality germplasm. Nevertheless, large-scale propagation of the developed cross-bred variety in the smallholder and backyard production system can yield higher productivity and livelihood sustainability of pig farmers (Kadirvel et al. 2023). Capacity building and training of pig farmers on scientific pig production practices is another key area which need to be undertaken on a wide scale (Bharati et al. 2022). This is very much essential to eliminate scavenging system of pig rearing, still prevalent in the country. A trained pig farmer can better manage pig breeding, feeding, diseases and housing conditions and equips him with good decision-making ability which ultimately leads to the increase in profitability from pig farming, in any type of production system. Credit support on procurement of pigs, pig feed, building of pig farm and contingency can support motivated youths and traditional pig farmers to undertake pig farming on a commercial level. Active veterinary support in terms of preventive health-check, farrowing assistance, vaccination, disease diagnosis and treatment are crucial to farmers. The availability of government veterinary clinics and hospitals, mobile dispensaries and emergency facilities are key to maintaining healthy herd (Bharati et al. 2022). Nevertheless, the availability of institutionally supported specially dedicated slaughterhouse for pigs will result in hygienic pork production, resulting into greater market returns (García-Díez et al. 2023). Another key area which requires technological intervention is piggery waste management. The rapid expansion of pig farms in the near future is going to pose a substantial pig farm generated waste management challenge (Lalthlansanga et al. 2023). Adoption of integrated pig farming practices can substantially reduce this problem, nevertheless, there exists constraints of land and space availability for such system to be widely adopted. Hence, the available technologies for piggery waste biodegradation and recycling need to be popularized and integrated into semi-intensive and intensive pig rearing system to reduce the environmental foot prints of pig rearing (Yang et al. 2023). Another area is development of feed compounding technologies for substitution of commercial pig feed with locally available non-conventional feed resources so that the input cost of pig rearing can be decreased. With this the innovation and development in technologies for precision feeding in the pig is also required. Precision feeding would warranty the delivery of the correct amount of feed with the accurate nutritional composition to individual animal when required at the exact time (Pomar et al. 2019). Thus, the nutrients in feed are tailored to meet the real-time requirements of the animal, thus increasing the feeding efficiency and feed conversion ratio. The development of technology intensive hygienic pork processing chain and its integration with digital technologies will primarily lead to high safety standards pork and pork products which is crucial to the consumers and addresses public health concerns (Johler and Guldimann 2021).

11.12 Conclusion

Piggery sector in India can become a budding economic enterprise and transform into a highly successful industry with enormous potential for sustaining livelihood and nutritional security. A combination of good managemental practice, easy credit accessibility, market linkage in conjugation with skill development of farmers through training programmes, can make any system of pig farming a wholesome profitable business. With the right support from the government, financial institutions, research organizations and private sector partners, commercial pig farming has great potential to contribute significantly to India's agricultural and economic growth.

References

AHIDF (2020) Implementation guidelines 2.0 for animal husbandry infrastructure development fund. https://ahidf.udyamimitra.in/pdf/animal-husbandry-infrastructure-development-fund.pdf

Alarcón LV, Allepuz A, Mateu E (2021) Biosecurity in pig farms: a review. Porcine Health Manag 7(1):5

BAHS (2023) 20th Livestock census—All India Report, Ministry of Agriculture Department of Animal Husbandry, Dairying and Fisheries, New Delhi. https://dahd.nic.in/sites/default/filess/BasicAnimalHusbandryStatistics2023.pdf

Bhaduria P, Satbir S, Singh A, Inderjeet, Parvender S (2023) Pig farming: techniques & technologies. ICAR-ATARI, Zone-1, Ludhiana, p 125

Bharati J, De K, Paul S, Kumar S, Yadav AK, Doley J, Mohan NH, Das BC (2022) Mobilizing pig resources for capacity development and livelihood security. In: Agriculture, livestock production and aquaculture: advances for smallholder farming systems, vol 2. Springer, Cham, pp 219–242

Bharati J, Kumar S, Mohan NH, Das BC, Devi SJ, Gupta VK (2023a) Ovarian follicle transcriptome dynamics reveals enrichment of immune system process during transition from small to large follicles in cyclic Indian Ghoongroo pigs. J Reprod Immunol 160:104164

Bharati J, Kumar S, Kumar S, Mohan NH, Islam R, Pegu SR, Banik S, Das BC, Borah S, Sarkar M (2023b) Androgen receptor gene deficiency results in the reduction of steroidogenic potential in porcine luteal cells. Anim Biotechnol 34(7):2183–2196

Bora M, Bora DP, Manu M, Barman NN, Dutta LJ, Kumar PP, Poovathikkal S, Suresh KP, Nimmanapalli R (2020) Assessment of risk factors of African swine fever in India: perspectives on future outbreaks and control strategies. Pathogens 9(12):1044

DAHD (2022) NAP on pig. https://dahd.nic.in/sites/default/filess/NAP%20on%20Pig%20.pdf

Deb R, Sonowal J, Sengar GS, Pegu SR, Praharaj MR, Malla WA, Singh I, Yadav AK, Rajkhowa S, Das PJ, Bharati J (2022) Porcine circovirus type 2 infected myocardial tissue transcriptome signature. Gene 836:146670

García-Díez J, Saraiva S, Moura D, Grispoldi L, Cenci-Goga BT, Saraiva C (2023) The importance of the slaughterhouse in surveilling animal and public health: a systematic review. Vet Sci 10(2):167

IS 3916:1966 (1966) Code of practice for pig housing. Bureau of Indian Standards. https://www.services.bis.gov.in/php/BIS_2.0/bisconnect/knowyourstandards/Indian_standards/isdetails/

Johler S, Guldimann C (2021) An introduction to current trends in meat microbiology and hygiene. Curr Clin Microbiol Rep:1–5

Kadirvel G, Kumaresan A, Das A, Bujarbaruah KM, Venkatasubramanian V, Ngachan SV (2013) Artificial insemination of pigs reared under smallholder production system in northeastern

India: success rate, genetic improvement, and monetary benefit. Trop Anim Health Prod 45:679–686

Kadirvel G, Devi YS, Naskar S, Banik S, Singh NS, Gonmei C (2023) Performance of crossbred pigs with indigenous and Hampshire inheritance under a smallholder production system in the eastern Himalayan hill region. Front Genet 14:1042554

Kumar S, Bhushan B, Kumar A, Panigrahi M, Bharati J, Kumari S, Kaiho K, Banik S, Karthikeyan A, Chaudhary R, Gaur GK (2023) Elucidation of novel SNPs affecting immune response to classical swine fever vaccination in pigs using immunogenomics approach. Vet Res Commun 48:1–13

Lalthlansanga C, Pottipati S, Meesala NS, Mohanty B, Kalamdhad AS (2023) Evaluating the potential of biodegradation of swine manure through rotary drum composting utilizing different bulking agents. Bioresour Technol 388:129751

Maes DG, Dewulf J, Piñeiro C, Edwards S, Kyriazakis I (2020) A critical reflection on intensive pork production with an emphasis on animal health and welfare. J Anim Sci 98(Supplement_1):S15–S26

Mohan NH, Pathak P, Buragohain L, Deka J, Bharati J, Das AK, Thomas R, Singh R, Sarma DK, Gupta VK, Das BC (2023) Comparative muscle transcriptome of Mali and Hampshire breeds of pigs: a preliminary study. Anim Biotechnol 34(8):3946–3961

NCDC (2017) Detailed project report on setting up of a pig rearing farm. https://www.ncdc.in/documents/downloads/171804052017.Sample-DPR_Pig-Rearing-Farm.pdf

Nhu TT, Dewulf J, Serruys P, Huysveld S, Nguyen CV, Sorgeloos P, Schaubroeck T (2015) Resource usage of integrated pig–biogas–fish system: partitioning and substitution within attributional life cycle assessment. Resour Conserv Recycl 102:27–38

Pomar C, Van Milgen J, Remus A (2019) 18: Precision livestock feeding, principle and practice. In: Poultry and pig nutrition: challenges of the 21st century. Wageningen Academic Publishers, Wageningen, pp 89–95

Rai A, Sharma SD, Sahoo PM, Malhotra PK (2008) Development of livelihood index for different agro-climatic zones of India. Agric Econ Res Rev 21(2):173–182

Sahu K (2022) A study about the piggery sector sustainability, pig breed diversity and its ecological implications in urban districts of Uttarakhand. Biol Life Sci Forum 15(1):19

Singh M, Pongenere N, Mollier RT, Patton RN, Yadav R, Katiyar R, Jaiswal P, Bhattacharjee M, Kalita H, Mishra VK (2023) Participatory assessment of management and biosecurity practices of smallholder pig farms in North East India. Front Vet Sci 10:1196955

Tripathi SD, Sharma BK (2001) Integrated fish-pig farming in India. FAO Fisheries technical paper. pp 54–56

Yang P, Yu M, Ma X, Deng D (2023) Carbon footprint of the pork product chain and recent advancements in mitigation strategies. Food Secur 12(23):4203

Entrepreneur Development Through Pig Farming

S. Swetha Kanthi and Srikanth Vallabaneni

Abstract

Entrepreneurs aspiring to explore diverse business opportunities within pig farming, along with its subsidiary ventures, encounter a unique dilemma influenced by social class, social order, and religion-influenced market segmentation on one side, and the rising trajectory of the pork market, uptrend of economic promise stemming from the untapped potential of the piggery industry on the other side. The crossroads between the merits and demerits of pig-based enterprises need to be understood before venturing into it. The chapter aims to discuss commercial entrepreneur development through pig/swine farming and associated industries through rationally reviewed piggery-based ventures. The strategies for sustainability and governmental initiatives aimed at nurturing the progress of the pig farming-based industry are outlined. The information furnished is intended to work as a self-evaluation tool for knowledge and business understanding to facilitate a nuanced venturing individual aiming to establish a holistic piggery-based enterprise from scratch. Futuristic pork markets and successful startups are discussed to arrive at isotones of successful piggery formulas. Training and skill development along with economic theories compliance of piggery-based enterprise are described for insights.

S. S. Kanthi (✉)
Department of Veterinary & A.H Extension Education, Sri Venkateswara Veterinary University, College of Veterinary Science-Proddatur, Proddatur, Andhra Pradesh, India

S. Vallabaneni
Division of Livestock Production & Management, Indian Veterinary Research Institute, Izatnagar, Izatnagar, Bareilly, Uttar Pradesh, India

© The Author(s), under exclusive license to Springer Nature Singapore Pte Ltd. 2024
T. Rana, B. Soto-Blanco (eds.), *Good Practices and Principles in Pig Farming*, Livestock Diseases and Management,
https://doi.org/10.1007/978-981-97-4665-1_12

Key.words

Global consumption status of pork · Pig-based startups · Challenges and opportunities · Routine piggery operations · Marketing of pigs · Executive summary · DPR · Futuristic ideas for pig-based enterprises

12.1 Scope of Entrepreneurial Opportunities Through Pig Farming

Despite being the fifth-largest producer of meat, exporting 894.04 MT of pig products worth Rs.186 million during the year 2020–21 (APEDA 2023), India faces a significant challenge and opportunity in the fact that its per capita pork consumption ranks 150th in the world. This underscores a critical area for both concern and potential entrepreneurial growth. The famous quote though a red herring attributed to Winston Churchill "Dogs look up to man. Cats look down on men. Give me a pig!" Indeed is a scientifically and economically rationalized fact that pigs have the potential to lift the standard of men involved in farming through the new generation of piggery-based enterprises. The entrepreneurial opportunity lies in the commercial aspect of the pig's breeding prolificacy. An adult female that starts breeding as early as 6 months of age delivers piglets twice a year with an average fecundity of around 8–12 piglets/delivery, these piglets weigh around 1 kg each at birth, and in less than a commercial year they grow to 100 kg. A growth of 100 times! Which is not seen in any other business forms. Apart from that, pork meat is highly nutritious and ranked as tasty. The omnivorous nature of the pig is the economic ease factor that allows the entrepreneur to choose from a variety of feeds and aim for profits. There are no hard and fast rules for feeding a pig which makes it easy for the entrepreneur to plan the feeding schedules with the locally available surplus and seasonal foods without worrying about the lean. In many countries, they are raised on food industry waste and dairy farm waste, they are considered Mortgage lifters as they can utilize excess milk and whey. Pig farming is the best mixed-farming choice for a fish farm. While India exports pork meat it also imports processed pork products at the same time, an estimated 500 metric tons of pork was imported during the year 2018. And this is the gap and opportunity an entrepreneur can cash in.

The majority of the pig production in India is recorded in the Northeastern States and now the commercial semi-intensive farms are registered in the states of Haryana, Kerala, Punjab, Andhra Pradesh, and a few other states too. Pig is recognized as an intelligent and insightful creature, displaying notable levels of intelligence and social behavior. Typically non-aggressive, it tends to exhibit minimal hostility unless provoked, especially when acting on its maternal instincts. Contrary to the common belief that associates pigs with dirtiness due to their omnivorous nature of feeding, many are unaware of the meticulous hygiene habits of pigs. They are, in fact, exceptionally clean animals, often designating specific areas for defecating, even piglets follow this soon after their birth. The Large White, Duroc, Landrace, Pietrain, and Hampshire are the five top adaptable global breeds of pig. However,

Table 12.1 Global per capita pork consumption and popular pig breeds by country

Rank	Country	Per capita consumption/year (kg)	Popular breeds
1	China	55.24	Meishan, Bama, and Wuzhishan, Taihu, Beijing black (domestic pig)
2	Poland	54.95	Polish landrace, Large white, pulavsaka (local primitive breed)
3	Spain	52.56	Duroc, Pietrain, Landrace, Iberico (native unique breed found only in Spain)
4	Lithuania	50.69	Landrace, Lithuanian pig, Heat-tolerant pig
5	Croatia	49.63	Turopolje, Black Slavonian pig
6	Hungary	48.3	Mangalica (only wooly pig breed in the world and the world's fattest breed)

across the world, nearly 400 breeds have been exploited, with the largest number of breeds being found in Asia and Europe (Groeneveld et al. 2010).

12.2 Global Status of Pork Consumption

Pork constitutes an estimated 36% of the global meat consumption, with 112.6 kilotons of consumption per year, the projected consumption by 2031 is 129 kilotons. Though India consumed over 292 thousand metric tons of pork during the year 2023, with 0.24 g of per capita consumption India's ranking is insignificant in the world pork per capita. Low per capita is the bull's eye of opportunity and holds untapped potential still in the pork industry. There is a bundle of room for the expansion of an entrepreneur through pork markets as the global market is open for the export of meat after liberalization (Table 12.1).

12.3 Basic Glossary

Entrepreneur   A person who starts a business and is willing to risk loss to make money.

Enterprise   An organization, especially a business, or a difficult and important plan, especially one that will earn money.

Entrepreneurship   Refers to an individual or a small group of partners who strike out on an original path to create a new business.

Pig entrepreneur   An entrepreneur who is into the business of making money from an industry that sells pork and pork products or associated subsidiary business.

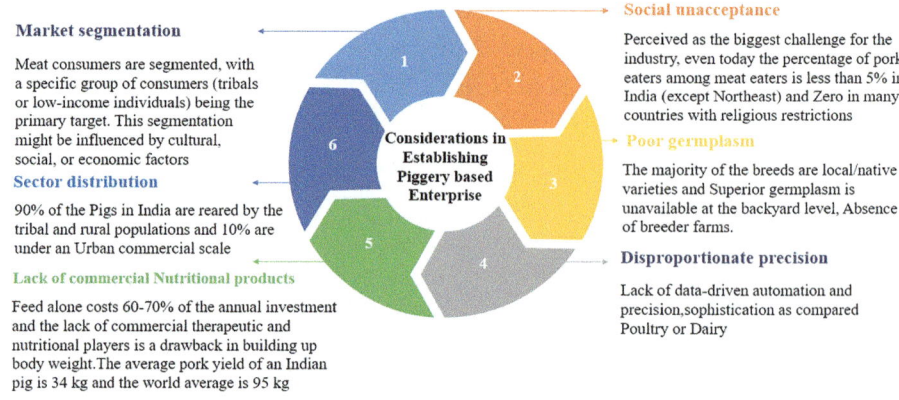

Fig. 12.1 Considerations in establishing piggery-based enterprise

12.4 Considerations in Establishing Piggery-Based Enterprise

See Fig. 12.1.

12.5 Comparing Challenges and Opportunities in Piggery

While talking about pig entrepreneurship it would be convenient to study the scenarios in the past present and future for deeper insight for an evidential change in the sector. Market liberalization, Awareness, Government support, Globalization, and Accessibility act as markers (Table 12.2).

12.6 Entrepreneur Development

For an entrepreneur planning to venture into a pig farming enterprise, it is imperative to arrive at answers for the following:

Where should I start a farm?
What are the farm aspects I should be aware of?
Where can I refine my knowledge and get skill training?
Whom should I contact for breeding stock?
How can I approach the bank or Government for support?

Table 12.2 Comparing challenges and opportunities in piggery

Trait	Past entrepreneurs	Present entrepreneurs	Future scope for entrepreneurs
Investments	Never imagined piggery as a business opportunity	Realized the potential of the market but still lagging in finding the customer base	Seed funding and a startup support system, on par with other sectors
Business type	Mostly tribal and backward communities owned business	Small- to medium-scale farms are venturing. Heavy commercial scale is still unregistered	Large-scale production with processing facilities
Swine holding	Backyard or unclaimed animals were slaughtered and consumed as food	Medium holding without processing technologies	Large capacity holding with processing and packaging technologies
Infrastructure	No infrastructure and extra facilities provided	Adequate infrastructure is available but lack of modernization and technology	Huge investments in infrastructure and precision
Position	As piggery is associated with taboos this was never looked upon as a respectable choice of money-making	Newly opened market and the existing entrepreneurs are pioneers who are setting up the path for future entrepreneurs	Healthy competition and giants in the industry to build economic progress and improve pork production
Outreach	Pork was never traded out of communities	Trade has started and going on a progressive stage, still dependent on imported meat. Only 2% of pork is being processed	Total pork needs are met by internal production and the ability to export pork and processed meat
Individual identity	The pig rearer was never accepted as a common member of society	Acceptance of the large farm owners as entrepreneurs	Animal farm status on par with other livestock
Environmental concerns	Never was a factor	Strict regulations abided by law and resources	More vigilant environmental clearance and sustainability regulations
Technology	Age-old practices and backyard farming	Semi-intensive and intensive modes of farms are finding scope	Large-scale commercial precision farms with full automation
Market demand	Zero to least demand	Demand-based and market-based production mode	Visioned as a substitute for meat and high-demand commodity

(continued)

Table 12.2 (continued)

Trait	Past entrepreneurs	Present entrepreneurs	Future scope for entrepreneurs
Gender	Mostly female-managed backyard piggery	Only a handful of successful female entrepreneurs is available as of now	More women coming forward and equal opportunities for female entrepreneurs
Taxation	Nil	Taxation is as per general rules but agriculture produces receive subsidies and exemptions up to a certain extent	Tax liberation to promote growth and more taxation on natural resources used

12.7 Prerequisites, Market Study, and Site Preparation

A thorough market analysis is a prerequisite for the successful establishment of a pork enterprise. An entrepreneur can achieve this by identifying the target market and comprehending its demographics, an evaluation of the current and potential future demand for pork products will provide an outline for the product range to be produced. Examining the competitive landscape by discerning the strengths, weaknesses, and strategies of both current and potential competitors will provide a SWOT analysis for decision-making. A thorough prior assessment of the reliability and availability of suppliers and distribution channels for both inputs and products produced in the enterprise will ease marketing and establish channels for the pork market. By conducting direct market research surveys to personally gather feedback from potential consumers an entrepreneur can gain first-hand information. Getting associated with existing pig farm entrepreneurs of the locality, local farmers under small-scale production, and stakeholders, will contemplate the feasibility of the planned enterprise. Finally being attuned to environmental and ethical considerations, addressing community concerns in the area is of prime importance while considering the piggery-based enterprise in or around human dwellings, as piggery-based industries are still not welcoming the site for the enterprise has to be planned away from the human dwellings.

A thorough Detailed Project report for the construction of a pig farm is necessary and can be prepared by an expert based on the animal holding capacity planned as per budget. A basic DPR (Detail Bankable Project Report) for ten animal holdings will provide the economics and commercial factors involved.

12.8 Piggery Detail Bankable Project Report (DPR)

General assumptions made

(a) Ten breeding sows are kept as a constant breeding stock. Breeding accounts for about 100%. After eight or nine farrowing, acquired animals may replace the parent stock.
(b) An insurance claim will be made to cover death in breeding stock (not adult stock).
(c) Nine crops of piglets will be sold out in 60 months because the project is expected to last 5 years.
(d) After 2 months, the mother weans the pups, and half of each litter is raised to market age that is, 8 months before being sold. At weaning, the remaining half of the litter is gone.
(e) The price increase for meat or live animals would offset the incremental rate of increase in input costs.
(f) A 5-year planning horizon would be anticipated for bankability purposes.
(g) The plan is for the domestic meat market which is expected to generate revenue.

12.9 General Aspects

See Table 12.3.

12.10 Standard Economic Norms

See Table 12.4.

Table 12.3 General aspects of a pig project report

S.N.	Particulars	Value
1.	No. of sows (6–7 months old)	10
2.	No. of boars	1
3.	No. of batches	2
4.	The interval between two batches (months)	3
5.	No. of farrowing per year	2
6.	No. of piglets per sow per farrowing	11
7.	Mortality among piglets (weaners)	20%
8.	Mortality among fatteners	10%
9.	Mortality among adults (not considered for insurance)	
10.	Weaning period (months)	2
11.	Space requirement (sq. ft.)	
	(a) Boar	70
	(b) Lactating sow with its piglets	100
	(c) Dry sows	20
	(d) Fatteners of 3–5 months age	10
	(e) Fatteners of 6–8 months age	15

Table 12.4 Standard economic norms

S.N.	Particulars	Value
1.	Total area required for building (sq. ft.)	1090
2.	Storeroom (sq. ft.)	100
3.	Concentrate feed requirement (kg/day)	
	(a) Breeding stock	2.5
	(b) Weaner	0.3
	(c) Fattener (3–5 months age)	1.5
	Total area required for building (sq. ft.)	1090
4.	Green fodder (kg/day/adult)	3
5.	Cost of construction of sheds (Rs./sq. ft.)	300
6.	Cost of construction of store room (Rs./sq. ft.)	500
7.	Cost of boar (Rs.)	6000
8.	Cost of sow (Rs.)	5000
9.	Cost of concentrate feed (Rs./kg)	10
10.	Cost of green fodder (Rs./kg)	1.5
11.	Insurance (%)	6
12.	Cost of medicines and vaccines (Rs./pig)	100
13.	Miscellaneous cost {power, water, other misc. expenses (Rs./pig)}	150
14.	No. of laborers required	1
15.	Laborers' wages (Rs./month)	3000
16.	No of piglets sold per sow per farrowing (2 months old)	4
17.	No. of fatteners sold per sow per farrowing (8 months old)	4
18.	Sale price of piglet (Rs./piglet)	2000
19.	Avg. wt. of fattener (kg)	80
20.	Sale price of fattener/parent stock (Rs./kg)	70
21.	Closing stock value of parents	150% of the original value
22.	Income from manure	
	Weaner/fattener (Rs./month)	205
	Adults (Rs./month)	50
23.	No. of gunny bags per ton of feed	20
24.	Sale of gunny bags (Rs./bag)	10
25.	Depreciation on sheds (%)	5
26.	Margin money (%)	15
27.	Interest rate (%)	12
28.	Repayment period (years)	5
29.	Grace period (years)	1

Table 12.5 Economic analysis of the project

	Details of investment	Indian currency-INR
A	*Capital cost*	
	Pig sheds and other structures	377,000
	Farrowing pens (4) for lactating sow	120,000
	Boar cum service pen	21,000
	Dry sow pens (6)	36,000
	Fattener shed-I (20)	60,000
	Fattener shed-II (20)	90,000
	Storeroom	50,000
	Water supply system	20,000
	Breeding stock: Crossbred gilts 10 Nos.	
	@ Rs. 5000 each and Crossbred boar 1 No.	
	@ Rs. 6000 each	56,000
	Total capital cost	**453,000**
B	*Recurring cost*	
	Capitalized recurring expenses up to 2 months	201,940
C	*Total project cost*	
	Capital cost	453,000
	Recurring cost capitalized	201,940
	Total capital cost	654,940
	Margin money (15% of total capital cost)	98,241
	Net finance required from the bank	556,700

12.11 Economic Analysis of the Project

See Table 12.5.

12.12 Annual Expenses Plan

See Table 12.6.

12.13 Income

See Table 12.7.

Total net income in 5 years: Rs. 457,378
Net income per annum: Rs. 91,476
Net income per month: Rs. 7623

Table 12.6 Annual expenses plan

S.N.	Particulars	I year	II year	III year	IV year	V year
A	**Fixed cost**	**453,000**				
B	**Recurring cost**					
1.	(a) Breeder feed cost					
	Concentrate	88,500	100,375	100,375	100,375	100,375
	Green fodder	15,930	18,068	18,068	18,068	18,068
	(b) Weaner feed cost	14,400	28,800	28,800	28,800	28,800
	(c) I-Fattener feed cost	27,000	144,000	81,000	81,000	81,000
	(d) II-Fattener feed cost		81,000	144,000	144,000	144,000
2.	Insurance cost (6%)	3360	3360	3360	3360	3360
3.	Labor charges	36,000	36,000	36,000	36,000	36,000
4.	Medicines	12,100	23,100	23,100	23,100	23,100
5.	Misc. expenses	4650	19,650	19,650	19,650	19,650
	Total	**201,940**	**454,353**	**454,353**	**454,353**	**454,353**

Table 12.7 Annual returns

S.N.	Particulars	I year	II year	III year	IV year	V year
1.	Sale of piglets @ Rs. 2000	80,000	160,000	160,000	160,000	160,000
2.	Sale of fatteners (Rs. 70 × 80 kg)		448,000	448,000	448,000	448,000
3.	Sale of gunny bags (260)	2600	7080	7080	7080	7080
4.	Sale of manure	7050	16,200	16,200	16,200	16,200
	Total	89,650	631,280	631,280	631,280	631,280
Cash flow analysis						
	Cost	*I year*	*II year*	*III year*	*IV year*	*V year*
1.	Capital cost	453,000				
2.	Recurring cost	201,940	454,353	454,353	454,353	454,353
3.	The depreciated value of the shed	0	0	0	0	299,730
4.	Closing stock value of animals	0	0	0	0	84,000
	Total	89,650	631,280	631,280	631,280	1,015,010
5.	**Profit**	112,290	176,927	176,927	176,927	560,657
Repayment schedule						
1.	Income	89,650	631,280	631,280	631,280	1,015,010
2.	Expenditure	Loan	454,353	454,353	454,353	454,353
3.	Gross surplus	89,650	176,927	176,927	176,927	560,657
4.	Loan balance	556,700	417,525	278,350	139,175	0
5.	Equated annual instalment	0	139,175	139,175	139,175	139,175
6.	Interest	66,804	50,103	33,402	16,701	0
7.	Net surplus	22,846	−12,351	4350	21,051	421,482

12.14 Economic Evaluation of the Project

An economic evaluation of the present project is carried out using net present value (NPV), internal rate of return (IRR), and benefit-cost (B/C) ratio. The prevailing bank rate (@12%) is taken as the discounting factor to calculate the present value and IRR.

Net present value (NPV): Rs. 53,292.45
Internal rate of return IRR): 166%
Benefit-cost ratio (B/C): 1.21:1

12.15 Economic/Financial Viability

The total initial fixed cost for a project involving ten sows is Rs. 453,000. When considering the recurring cost for the first 12 months as a capitalized cost, the overall capital requirement for the project would amount to Rs. 654,940. With 15% of the total cost allocated as margin money, a net bank loan of Rs. 556,700 is needed. Over a 5-year planning horizon, the project is projected to yield net returns of Rs. 457,378, resulting in a net income of Rs. 91,476 per annum and Rs. 7623 per month.

The net present value (NPV) of the project stands at Rs. 53,292.45, reflecting its high economic viability. Additionally, the internal rate of return (IRR) is an impressive 166%, further affirming the economic and financial robustness of the project. This indicates that the scheme is not only technically feasible but also economically sound, making it a viable venture.

12.16 Layout Plan for 200 Pig Holding

While the DPR for any number of animal holding is arrived at by multiplying the basic ten animal holding DPR, the floor spaces and plan may slightly vary based on sophistication and additional facilities. A basic layout for 200 pig holding is shown here, each block of 50 piglets will have the following layout so an overall four similar blocks for the 200 piglets holding would be required (Fig. 12.2).

12.17 Noncompliance of Economic Theories with Piggery

With its unique variant feature of dealing with live animals and dealing with biological fluctuations, the very nature of animal-based enterprise doesn't comply with the commodity-based business or fintech standards. Several economic theories find less scope or modified scope while talking about pig farming enterprise. The application of economics in animal-based enterprises has new dimensions. Following are ten considerable economic theories that don't sit fit with piggery-based enterprise:

1. **Classical economics:** Classical economics, associated with economists like Adam Smith, emphasizes the concept of rational self-interest. According to this theory, individuals act in a way that maximizes their utility or satisfaction. If a purchase is perceived as beneficial to the individual, they are more likely to make that purchase. But in the pig or pig-based industry even though the cus-

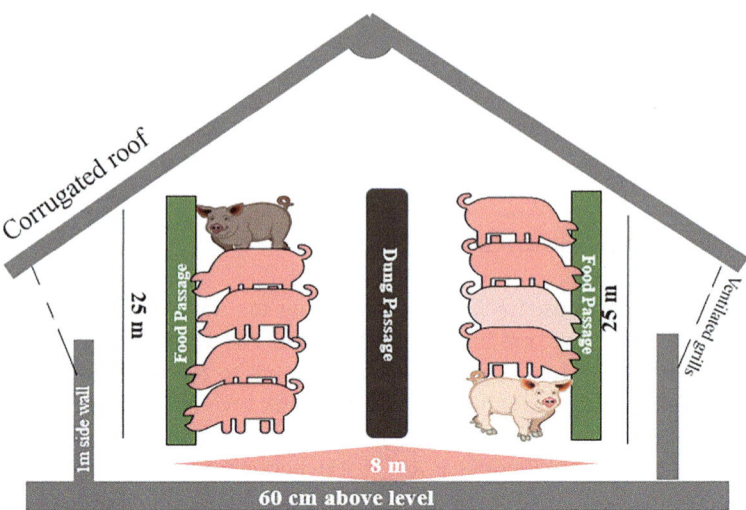

Fig. 12.2 Layout plan for 50 pig-holding sty

tomer perceives the product to be useful the purchase is not made because of the underlying behavioral and religious aspects.
2. **Behavioral economics:** Behavioral economics incorporates insights from psychology into economic analysis. It recognizes that individuals may not always act rationally and can be influenced by psychological factors and beliefs. Piggery is highly influenced by behavior of the society. The concept of "loss aversion," where individuals place a higher value on avoiding losses than acquiring equivalent gains, might be relevant. In this case, individuals may view violating religious dietary restrictions as a loss, influencing their decision-making.
3. **Herd behavior**: Herd behavior analyses that the customer mimics the larger group purchasing trend which is not so with the constant pig enterprise market where sudden fluctuations or sudden market shifts are not observed. Immunity to shift towards herd behavior is predominant towards pork and pork-based products.
4. **Labor theory of value:** Marxian economics argues that the value of a service or good is based on the socially necessary labor involved in production. In pig enterprise, the value of an animal or product produced is dependent on factors beyond labor. Genetic makeup, reproductive stage of the animal, breed of the animal, market demand of the specific product, health conditions, etc. have greater influence.
5. **Supply and demand**: The equilibrium between supply and demand can never be met as per the market ask, due to the biological nature of pig farming the supply can be surplus at times and scarce which can never be adjusted to immediate fluctuations not complying with demand equilibrium models. An efficient market hypothesis can never be made in this context.
6. **Rational choice-making theory**: Non-monitory factors often play a great role in pig farming rather than a perfect decision-making process, a rational decision

can always be sidelined while considering personal prejudice, taboos, beliefs, culture, environmental factors, etc. Pig farming is not opted for even if it is a profitable venture enterprise, especially in the Indian sub-continent where pig is a religious and low social order obligation leaving limited scope for rational decision-making at every step.
7. **Marginal productivity**: The marginal productivity theory says that an increase in input units like labor or raw material will bring in output change. This holds good for any enterprise but not for animal production. Risk return trade-offs are not applicable in a pig enterprise.
8. **Purchasing power parity (PPP)**: The application of PPP with a basket of goods approach doesn't happen in the case of pig farming-based commodities. PPP states that when the frictions are constant/absent and goods are identical, then goods should be sold at the same prices in the absence of transportation costs. This means evaluating PPP in common currencies, Pork or Bacon produced in the USA and EU should give an exchange rate idea, but in reality, the cost of production is highly variable across nations for pig farming-based enterprises.
9. **Risk neutrality**: Risk is the synonym of an enterprise that deals with unpredictable animal behavior, ranging from climate, feed, water, variations in microenvironment, and physiology, etc., every factor influences the output of the pig enterprise making it imperfect for risk-neutral market. The time value of money and dividends can be not the same as in any other enterprise.
10. **Fisher exchange equation**: Pig enterprise cannot be scaled in terms of inflation and interest as both have the least effect on the sector. According to Fisher equation nominal and real interest rates are associated with inflation. It can be modified or limited for monetary evaluation. Food-based industries falling under essential commodities don't comply with it.

12.18 Understanding Aspects of Farm Operations

Acquiring a comprehensive understanding of pig farming-related skills and knowledge is essential to start a business related to piggery. To attain this proficiency, new entrepreneurs should seek training from experts in the area. Several ICAR-based and SAU-based government institutions and some NGO and Cooperative societies offer skill training and regular upgrade training to aspiring entrepreneurs. Given the complexity of pig farming, it is impractical for a novice entrepreneur to learn everything independently without hands-on experience. Seeking guidance and expertise from seasoned professionals can significantly enhance the entrepreneur's learning curve and contribute to the success and sustainability of the pig farming venture.

12.19 Periodic Farm Activities in a Pig Enterprise

See Table 12.8.

Table 12.8 Periodic activities in a pig farm

S. N.	Operation	Importance	Procedure
1.	Navel cord treatment	To prevent infections through the umbilicus	Cut the navel cord, leaving a 2.5–3.5 cm (1–1.5 inch) portion that needs to be soaked in a tincture iodine solution (7%)
2.	Needle teeth clipping	To prevent damage to the sow's udder or the fellow pen mates	Needle teeth (8 no's) can be cut using clippers or pliers as soon as possible after birth
3.	Iron supplementation	To shield piglets against nutritional anemia	Can be supplemented through iron dextran injection, ferrous sulfate salt paste applied to the sow's udder, or oral iron supplements
4.	Piglet identification	For easy tracking, care, and management of animals	This can be done using tattoos, ear tags, ear notching, etc.
5.	Colostrum feeding	To provide proper nutrition and also passive immunity to the piglets	Piglets should suckle colostrum immediately after birth either natural or artificial colostrum
6.	Weaning	Improves sow productive life	Separating piglets from suckling sow viz., conventional/split/early weaning/ early specialized weaning (segregated early weaning (SEW) & medicated early weaning (MEW))
7.	Fostering/ cross-fostering	To equalize the litter sizes in a farm or for rearing orphan/ disowned/at-risk piglets	Piglets are moved to a different sow than their mother to be nurtured and it can be straight/even, cross, back fostering or shift/split suckling
8.	Creep feeding	Improves digestion of feeds other than milk and promotes piglet growth	Feeding protein-rich, highly digestible solid diet to piglets from 2nd week of age
9.	Weighing	To monitor the growth rate and performance	Can be done weekly using weighing scales or through formulae (girth method)
10.	Heat detection	For proper breeding management	This can be done via physical, and chemical methods
11.	Breeding	For producing the next generations of pigs	Pen/herd/pasture/hand mating or artificial insemination
12.	Flushing	To increase the number of eggs produced and thereby improve the reproductive success rate	Providing sows and gilts with extra nutrition 1 to 2 weeks before mating
13.	Detection of pregnancy	To ensure better feeding and other management during pregnancy	This can be done through various methods viz., physical, chemical, biological, direct/indirect techniques

(continued)

Table 12.8 (continued)

S. N.	Operation	Importance	Procedure
14.	Castration	To reduce the boar taint and improve pork quality, improve growth rate, decrease aggression	Male piglets can be castrated after 1 week of age via surgical or by immunological methods
15.	Tail docking	To lessen pig cannibalism and tail-biting	Should be early before weaning using clippers/pliers or cauterizing tail docker
16.	Feeding and ration formulation	Pigs should be fed based on their weight and nutrient requirements	It can be creep/weaner/grower/finisher feed depending on the life stage of the animal
17.	Cleaning and disinfection of sty	To control bad odor and keep the animal house clean and tidy	Should be done using water and chemical disinfectants or detergents
18.	Treatment of sick animals	To improve the welfare of pigs by reducing pain and suffering	Timely treatment helps in improving the animal's condition thereby the farm's profitability
19.	Segregation/ isolation/ quarantine	To reduce the chances of disease spreading	Isolation and quarantine sheds should be constructed in the animal farm to separate diseased or new animals in the herd
20.	Culling	It has a direct bearing on how profitable the breeding herd is	Animals on the farm that are underproductive and unproductive should be removed from the herd promptly
21.	Deworming & vaccination	Improves the growth rate and avoids the occurrence of infection in the herd	Animals should be dewormed and vaccinated depending on the common diseases in the area according to the schedule
22.	Record keeping	The only way to determine a farm's final profit or loss is to maintain accurate records	Various records such as an animal data sheet, the total number of animals in the stock, a feed register, an expense statement, etc.

12.20 Important Goals to Accomplish in Pig Farm Management

See Table 12.9.

12.21 Standard Objectives or Targets for Production

See Table 12.10.

Table 12.9 Important goals to accomplish in pig farm management

1. Reproduction rate: 10 to 11 live piglets should be born to each sow in a litter.
2. A sow should give birth to at least two litters a year, or 20–22 live piglets per sow.
3. Survival rate: A minimum of 85% of live births should be raised to weaning (17–19 piglets per sow annually).
4. Growth rate: Aim for 90 kg of live weight in 170 days, and a lifetime carcass yield of 77%, or 500 g per day.
5. Aim for a 3.5 kg feed conversion efficiency.
6. At the age of 4–6 months, when they will have reached a market weight of 65–100 kg, aim to market the pigs for slaughter as porkers.
7. Make every effort to provide the sows with a maximum 5-year producing life span.

Table 12.10 Standard objectives or targets for production

1. Of the sows bred, 90% should become pregnant
2. Having a litter of at least 12 piglets
3. Piglets born weighing 1.3 kg or more
4. At weaning, piglets weighing 11 kg or more
5. After weaning, a daily weight gain of at least 0.5 kg

12.22 Institutional Support

An individual novice entrepreneur requires institutional support at several stages. The superior germplasm can be purchased from

ICAR RC for NEH, Nagaland Centre, Medziphema
Assam Agricultural University, Khanapara
State Veterinary Department, Aizawl, Mizoram
State Animal Husbandry and Veterinary Department, Govt. of Arunachal Pradesh
Animal Resource Development Department, Govt. of Tripura, Agartala, Tripura
Birsa Agricultural University, Ranchi
Chhattisgarh Kamdhenu Vishwavidyalaya, Durg, Chhattisgarh
Kerala Veterinary and Animal Science University, Pookode, Kerala

Regional All India Coordinated Research projects on pigs support regional entrepreneurs in providing superior germplasm:

1. College of Veterinary Science, Assam Agricultural University, Khanapara, Guwahati, Assam
2. College of Veterinary & Animal Science, Kerala Veterinary and Animal Science University, Mannuthy, Kerala

3. College of Veterinary Science, Sri Venkateshwara Vety. University, Tirupati
4. ICAR-Central Coastal Agricultural Research Institute, Ela, Old Goa
5. Indian Veterinary Research Institute, Izatnagar, U.P.
6. Tamil Nadu Veterinary and Animal Sciences University, Kattupakkam, Tamilnadu
7. College of Veterinary Science & AH, CAU, Selesih, Aizawl, Mizoram
8. SASARD, Medziphema, Nagaland
9. ICAR-Central Island Agricultural Research Institute, Garacharama, Port Blair, Andaman and Nicobar Islands
10. College of Agriculture, CAU, Iroisemba, Imphal, Manipur
11. ICAR Research Complex for NEH Region, Umroi Road, Umiam, Meghalaya
12. Indian Veterinary Research Institute, Eastern Regional Station (ERS), 37, Kolkata
13. KVK Goalpara, Dudhnoi, ICAR-NRC on Pig, Rani, Guwahati, Assam
14. Krantisinh Nana Patil College of Veterinary Science (Maharashtra Animal and Fishery Sciences University), Shirval, Dist.- Satara, Maharashtra
15. College of Veterinary Science, Guru Angad Dev Veterinary and Animal Sciences University, Ludhiana, Punjab

Apart from these training and incubation centers of ICAR and SAU-based KVKs can be contacted which conduct skill and knowledge training regularly.

12.23 Financial Support

Government schemes provide subsidies to various entrepreneurs under different schemes, apart from this several agricultural, cooperative, and rural banks can be approached for financial assistance on a loan basis.

12.24 Facilities for Credit

The basic purpose of Governments providing support can be understood through Keynesian economics, developed by John Maynard Keynes, which talks about the role of aggregate demand in the economy. Keynes formulates that government intervention, particularly through fiscal policy, could stimulate demand and boost economic activity. In times of economic downturn, encouraging people to spend by making purchases beneficial through policies like tax cuts or public spending can help stimulate economic growth. In India, the following finance and insurance options are offered for the acquisition, marketing, and processing of pork.

Animal Husbandry Infrastructure Development Fund (AHIDF) The Animal Husbandry Infrastructure Development (AHIDF, 2023) has been approved for incentivizing investments by individual entrepreneurs, private companies, MSME, Farmers Producers Organizations (FPOs), and Section 8 companies to establish (i) the dairy processing and value addition infrastructure, (ii) meat processing and value addition infrastructure, and (iii) animal feed plant. Entrepreneurs can avail subsidies for pig breeding farms with modern infrastructure, pig fattening farms with advanced technology, and integrated production systems. Pig Entrepreneurs can also avail subsidies for polygon semen stations and artificial insemination technology for pigs.

National Bank for Agriculture and Rural Development (NABARD) The Pig farming-based Entrepreneurship Development Scheme (PEDS) and the Pig Development Scheme (PDS) are two of the credit options that the NABARD provides to pig farmers. Pig farmers can use these programs to get loans for marketing, operating capital, and infrastructure related to raising pigs. A detailed executive summary and DPR is required for bank proceedings.

Pradhan Mantri Mudra Yojana (PMMY) The government introduced the PMMY credit initiative to lend up to Rs. 10 lakhs to small and micro businesses, including those involved in the meat processing industry. The loan is given out by MFIs, NBFCs, and banks.

National Livestock Mission A centrally supported program called the National Livestock Mission (NLM) offers funding to support the growth of livestock, especially pigs. Reimbursements for pig-rearing infrastructure, including sheds, food supplies, and water facilities, are available to qualified farmers under the NLM.

Kisan Credit Card A credit facility known as the Kisan Credit Card (KCC) is provided by several financial organizations, such as banks and cooperatives. Pig farmers can use KCC loans to finance their pig farming endeavors, including buying piglets, feed, and equipment.

Livestock Insurance Scheme (LIS) This is a government-sponsored program that offers insurance protection against livestock deaths or losses brought on by illnesses, accidents, and natural disasters. Farmers that raise pigs have the option to use the LIS as a financial safety net against pig death.

Agricultural and Processed Food Products Export Development Authority (APEDA) APEDA (2023) offers financial support for the majority of its facilities. Grants, processing units for modernization, upgrades, expansion loans, or equity involvement are some of the ways that support might be provided.

Technology Upgradation Fund Scheme (TUFS) TUFS offers financial support to meat processing industry entrepreneurs so they can expand, upgrade, and modernize their facilities. The support comes in the shape of an interest rate subsidy on the loan that was obtained for the project.

The Infrastructure Development Fund is a welcoming opportunity to establish corridor businesses associated with piggery-based industries like Nutraceutical or Swine-specific pharmaceutical ventures, feed plants, SAS-based models, etc. Waste management and recycling piggery-based affluents are also supported under the scheme and have a wide scope.

12.25 Futuristic Business Ideas

Processing Industry It is reported that value addition of meat is limited and less than 2% of total meat is only processed into products for trade in India as compared to more than 60% in developed countries. A processing industry can be a fetching idea for entrepreneurs.

Nutraceutical Enterprise The potential of the untapped market for pig nutritional supplements is estimated to be around INR 12.35 billion. The industry currently suffers from a shortage of manufacturers, leading to challenges in pig farming such as inadequate bodybuilding, low weight gain, and reduced output. Establishing a feed manufacturing company in this sector presents a lucrative business opportunity.

Organic Pork Production The potential of organic pork production in tribal areas needs to be explored by government and non-government organizations. Commercial farms emphasizing organic meat will pull in a customer base who are under misconceptions about the feeding habits of pigs.

Pharmaceutical Enterprise The absence of swine-specific medication is identified as a contributing factor to low productivity and an inadequate response rate to treatment in the piggery-based industry. Introducing a labeled pharmaceutical production specifically designed for swine presents a futuristic and innovative concept that can fill the gap. This entails developing and manufacturing medications that are tailor-made to address the unique health needs and challenges faced by pigs. A labeled pharma production for swine could encompass a range of medications, including vaccines, antibiotics, and other therapeutic agents, formulated to enhance the overall health, productivity, and disease resistance of swine.

Service-Based Blockchain Traceability Firms Pig farming as understood is based on several misbeliefs and it is hard to convince the consumers regarding the meat quality and production. Implementing blockchain technology for supply chain traceability in the pork industry can enhance transparency, allowing consumers to trace the entire journey of pork products from farm to table, ensuring quality, safety, and ethical practices. Tech bases or SAS-based companies can look forward in this regard.

Service-Based Biosecurity Solutions Pig farming lacks precision and advances on par with other livestock sectors like dairy and poultry. Developing advanced biosecurity technologies which maintain biosafety using biometric technology, circular economy facilities, AI-powered surveillance, and disease detection algorithms to prevent and control the spread of diseases within pig farms, ensuring the health and safety of the herd relieves half the burden for the entrepreneur. The whole idea can be outsourced to many entrepreneurs.

Private Breeding Farms Entrepreneurs can capitalize on the absence of breeding farms specializing in superior pig germplasm, establishing ventures for significant profit. A robust breeding program, state-of-the-art facilities, genetic testing, and collaboration with research institutions are essential components. Continuous innovation and commitment to genetic improvement position these ventures for success in meeting the growing demand for high-quality pig germplasm.

12.26 Bankable Executive Summary for Piggery-Based Enterprise Execution and Marketing

A pig farming-based entrepreneur if planning to go for financial support from a bank for the initial setting up of the venture has to fulfill necessary documentation and financial procurement conditions as indicated, many third-party assistances are available in the market and can be availed if the entrepreneur is a novice and doesn't feel confident. A basic grasp of every facet of the enterprise equips one with insights and a firm hold on the dynamics occurring within the organization. Before preparing an executive summary for a bank loan or project approval detailed documentation is mandatory on the following aspects (Table 12.11).

12.27 Ten Successful Global Pig Farming-Based Startups

See Table 12.12.

Table 12.11 Executive summary of pig farming enterprise

Documents required	Components	Purpose/justification
Business idea form	Pig farmland/Procurement unit/Processing unit/Service provider/Feed unit/Value addition center/Machinery/Recycling unit etc.	Motivation to retain the business idea, what will the product be, who will be the target group, what is the pathway to reach the customers, what needs of customers will be met?
Market research form	Product(s) in hand, target customers, preferences, and purchasing criteria of the customers, existing gaps that competitors are unable to meet, longevity of the product, duplicates in the market, compliance with the acceptability norms of the target customer base	
Marketing plan form	Product and process	Plan for product design, color, packaging, size, certification, quality, process/steps of sales
	Price	Pricing of the product, how much customers are paying for the identical product, how much they are willing to pay for your product, reasons for the quoted price/justification, reasons and plans for discounts, and credit for customers
	Place	Reasons for selecting the specific location, monthly costs, and variables included in that, method of distribution (to local retailers, wholesalers, export etc.)
	Promotion/Promotion expenditure	Direct marketing, telemarketing, advertising, social media, publicity, sales marketing
	People	Position, recruiting criteria, plan for HR development, and external training
	Physical evidence	Infrastructure and signage, biosecurity, eco compliance
Annual sales estimate form	Direct and retail individual monthly plan, EBEITA	Total sales volume, total sales in target market, market share

(continued)

Table 12.11 (continued)

Documents required	Components	Purpose/justification
Enterprise structure form		Tasks and responsibilities, staffing, labor, designations, outsourced staff, marketing staff, part-time vendors, etc.
	HR expenditure and plan	Required skill set, responsibilities, pay as per skill sets, emoluments, appraisal criteria, welfare plan
Legal forms	Operating procedure and justification	Ownership, authority, and role description
Legal responsibilities	Taxes applied to the business, regulations for the staff, license and permits required, insurances applicable, eco laws followed	
Product costing form	Variable and fixed cost per item to arrive at the total cost of the item	Estimates on a monthly basis are calculated for the product portfolio separately for each one
Fixed cost form	Rent, electricity, machinery, transport, repair and marketing	
Depreciation form	Equipment, machinery, sheds	Estimated cost and life of each one described and depreciation estimated cost per year
Monthly purchase form	Estimated number of items sold per month as per market study	Variable cost and total cost of item
Sales plan	Sale volume, sales price, and sale value	Of all the portfolio products like pork, bacon, frozen meat, sausages etc.
Cost plan	Production volume, variable cost of each item, total variable cost	
Profit plan	Variable cost, fixed cost, gross outcomes, margins, EBEITA	
Cash flow plan		Annual estimate
Estimate of piggery enterprise		Capital required
Seed funding or angel funding sources, Govt. subsidies	Owners' equity, owners capital, required funding estimate, collaterals if going for a loan/applicable	
Loan repayment schedule	Amortized plans details	Plan to repay, funds available, collateral
Statutory requirements	Affluents clearance plan, CSR declaration, CSR is voluntary in the USA and mandatory in EU and India if covered under Sec. 135 India. GST and MSME registration if applicable, APEDA approvals if the product is exported, FSSAI compliance and clearance of the product, labor laws approval, fire and safety, pollution control board clearance especially for the affluents, odor, sound produced in a typical piggery-based industry	

Table 12.12 Successful startups in the piggery sector

S.N.	Name of the company	Geographical trade base	Unique feature/takebacks
1	Arohan Foods	India	Transforming the pork industry with Arohan's products, which are sold in 22 Indian states and include bacon, sausages, ham, and salami
2	Renthlei Piggery Farm	India	Mizoram state's high-quality piglet and pork-producing business model
3	Meatigo by Prasuma	India	B2C marketplace that sells meat and meat-related goods. Poultry, hog, bacon, kebab, mutton, seafood, and sausages in both raw and processed versions are among the product categories that it offers. It claims to offer premium farm-fresh food
4	Khaisua Foods	India	Utilizing retort technology to manufacture shelf-stable items for both offline and online sales, providing shelf-stable traditional pig meals from Assam
5	eGenesis	USA—Massachusetts-based biotech company	Successfully producing pigs with organs that could be safe enough to transplant into humans
6	SmartGuard	USA—Iowa	It combines voice recognition technology developed by humans with a wearable pad attached to the sow's flank. A slight vibration impulse alerts the sow, and the SmartGuard controller sends a signal to the pad when it hears a pig squeal. She releases the pig by shifting her weight
7	New Age Meats	USA—California	The first clean meat company to try producing sausages using pig's cells
8	E-Doctor	Singapore	A health-monitoring device that uses a wireless ear tag to gather real-time data on each pig's temperature, physical activity, and mating cycle to assist farmers in tracking the health and mating of their pigs
9	Zhenmeat	China	Introduced a vegan substitute for deep-fried pork and crayfish
10	Qihan Biotech	China	Gene-edited pigs for xenotransplantation

12.28 Markets and Marketing of Pork and Associated Products

Effective pig and pork marketing, alongside other production and management aspects, is vital for sustaining robust revenue in a piggery firm. The growing public demand for alternative protein sources fuels continual growth in the pork industry.

To ensure the sustainability of a pig processing business, understanding marketing dynamics and adhering to relevant rules in market selection are essential.

New-generation entrepreneurs need new-generation solutions for effective marketing. India is witnessing a digital revolution, this can be used for effective marketing of pork products for sustaining revenue in the growing pork industry, driven by increased public interest in alternative protein sources. A multifaceted modern approach includes establishing a robust online presence through websites and social media, integrating with e-commerce platforms for convenient online sales, for instance there are already several players in the market like Licious, Zpp fresh, Fresh Meat etc., and also locally it can be marketed through participating in farmers' markets and food festivals to directly engage with consumers. Tailoring marketing strategies to regional preferences, implementing subscription models, and leveraging influencers in health and fitness sectors further enhance market reach, subscription pork is going to be the future of the pork industry. Mobile apps, loyalty programs, quality packaging, and sustainability messaging also play crucial roles in creating a competitive edge and fostering customer loyalty. These innovative marketing methods can check the age-old critical issues in piggery.

References

Agricultural and Processed Food Products Export Development Authority (APEDA) (2023) Home. Retrieved from https://apeda.gov.in

Department of Animal Husbandry and Dairying, Government of India. 2023. Animal Husbandry Infrastructure Development Fund (AHIDF). https://dahd.nic.in/schemes/animal-husbandry-infrastructure-development-fund-ahidf.

Groeneveld LF, Lenstra JA, Eding H, Toro MÁ, Scherf BD, Pilling D et al (2010) Genetic diversity in farm animals—a review. Anim Genet 41(S1):6–31. https://doi.org/10.1111/j.1365-2052.2010.02038